The
EVERGLADES
HANDBOOK
Understanding the Ecosystem

Thomas E. Lodge

With an introduction by
Marjory Stoneman Douglas

St. Lucie Press

Front cover: A great white heron, a tropical representative in the Everglades, takes a golden shiner, a fish representing temperate North America. The scene depicts the two unifying themes of this book: *ecosystem function* represented by a wading bird eating a fish—probably the most important event in the ecosystem—and *biogeography,* showing the juxtaposition of tropical and temperate influences that characterize the plants and animals of the Everglades. (Photo by Robert Hamer.)
Back cover: Photo by Masud Quraishy.

Printed and bound in the U.S.A. Printed on acid-free paper.
10 9 8 7 6 5 4 3 2

Library of Congress Cataloging-in-Publication Data

Lodge, Thomas E., 1943–
    The Everglades handbook: understanding the ecosystem / by Thomas
E. Lodge : with an introduction by Marjory Stoneman Douglas.
      p. cm.
    Includes bibliographical references and index.
    ISBN 1-884015-06-9
    1. Swamp ecology—Florida—Everglades. 2. Swamp fauna—Florida—
Everglades. 3. Swamp flora—Florida—Everglades. 4. Everglades
(Fla.). 5. Ecosystem management—Florida—Everglades. I. Title.
QH105.F6L63 1994
574.5'26325—dc20   93-48083
                                                  CIP

    Direct all inquiries to St. Lucie Press, Inc., 100 E. Linton Blvd., Suite 403B, Delray Beach, Florida 33483.

Phone: (407) 274-9906
Fax: (407) 274-9927

S$\overset{t}{\text{L}}$

Published by
St. Lucie Press
100 E. Linton Blvd., Suite 403B
Delray Beach, FL 33483

Dedication

To the memory of

Frank C. Craighead, Sr. (1889–1982)
In retirement, he developed and shared
the deepest knowledge of the Everglades ever attained by one man

William E. Odum (1942–1991)
With a few others,
Bill put mangroves on the map and before the legislature

Oscar T. "Bud" Owre (1917–1990)
Mentor and friend to many students of the Everglades

Henry M. Truby (1919–1993)
Distinguished linguist, champion of dolphins, and friend

Edward Ellsworth Lodge (1906–1981)
From Silver Lake, Ohio, and finally to Florida,
with many enjoyable days in the Everglades: my father

Contents

Part IV. ENVIRONMENTAL IMPACTS

Acknowledgments

The initial inspiration for this book came from photographer Robert Hamer, who envisioned a joint project: a large-format book of his color pictures and my text. Masud Quraishy, a close friend of ours and owner of Kenya Photo Mural, Inc. of Miami, became involved due to his interest in promoting Bob's photographs. Bob and Masud both contributed considerable effort and resources to the larger concept of the project. Their support is appreciated, and the pictures in this book are black and white conversions by Kenya Photo Mural made from Bob's transparencies and negatives.

Many people provided technical reviews of various parts of the manuscript. Several faculty members of the University of Miami were of valuable assistance. Taylor R. Alexander, retired Professor of Botany, critiqued an early draft and was especially helpful concerning vegetation, drawing on his expertise in the Everglades which began in the early 1940s. Ronald H. Hofstetter, Professor of Biology, provided guidance on plant communities and the effects of fire. C. Richard Robins, Maytag Professor of Ichthyology, provided numerous comments concerning fishes, and Harold R. Wanless, Professor of Marine Geology, provided comments in the area of soils and geology. Oscar Owre, Maytag Professor of Ornithology Emeritus and a friend for many years, offered helpful comments on birds and on the final chapter. His untimely death is regretted, but his encouragement endures in these pages.

Reviews by the technical and administrative staffs of Everglades National Park were arranged by Michael V. Finley, then superintendent, whom I thank for his support. Of the park's research staff, special thanks are due William F. Loftus (for invertebrates, marine and freshwater fishes, and amphibians), John Ogden (who provided information on wading bird populations and a critique of the entire manuscript), Michael B. Robblee (information on Florida Bay), and William B. Robertson, who read and provided helpful comments on the entire first draft manuscript.

Other technical comments were provided by Norris Williams, Director of the

Florida State Museum (epiphytic plants); Lance Gunderson of the Department of Zoology, University of Florida (Chapters 2 and 3); Frank J. Mazzotti of the University of Florida, Institute of Food and Agricultural Sciences (reptiles); Richard W. Cantrell, Florida Department of Environmental Protection (invertebrates); and Paul Moler (crocodiles) and David S. Maehr (mammals, with particularly helpful information on the Florida panther), both of the Florida Game and Fresh Water Fish Commission. Thomas S. Baskett, U.S. Fish and Wildlife Service/ University of Missouri, read and provided helpful uncle's guidance on an early draft manuscript.

Comments from the South Florida Water Management District were made and coordinated by Cathleen C. Vogel. Other helpful comments came from David W. Black, Joseph B. Browder, Stephen W. Carney, Robert S. Carr, Richard B. Darling, Marjory Stoneman Douglas, Michael J. Duever, Scott G. Evanson, David J. Fall, Barry M. Faulkner, Angela R. Finney, Len Fishkin, Robert B. Halley, Robert L. Hamrick, Frances B. (my mother) and Herbert Heisterkamp, Robert W. Higgins, Hilburn O. Hillestad, Wes Jurgens (who also provided the generalized Everglades cross section on page 22), Robert L. Kelley, Karin Landers, Patricia H. Lodge, Laura Louthan, Michael D. Mesiano, MaryLynn Musgrove, Thomas A. Paris, Christine R. Pope, Nathaniel P. Reed, Peter C. Rosendahl, Paula Sessions, Linda L. Sheetz-Halley, Robert Silliman, the late Henry M. Truby, Dale H. Twachtmann, and James D. Webb. Lisa Holmes, Ellie Maysonet, Cheryl Mills, and Paula Serola provided typing support. Law Companies provided editorial assistance by Rhonda G. Vest, arranged by Mike Marshall, and graphics preparation by Reginald D. Dunnick and David J. Fall, with assistance by Manuel Encalada, George Kanakis, and Kenneth S. Slack. Other company guidance and help was provided by C. Edwin Copeland, Jr., John Ehrlichman, and Anne Ockene.

Special thanks are due Catherine H. Sweeney of Coconut Grove, Florida, whose help leading to this project began in 1972 when she provided me, then a graduate student, with a home and an introduction to Marjory Stoneman Douglas, and many others, amidst the history and excitement of "The Kampong" (see footnote on page 9). In that regard, I appreciate the support of Carolyn G. Abernathy and Harriet Fraunfelter in facilitating numerous connections.

And not to be forgotten is the tolerance of my wife, Patti, and my daughter, Morinda, during the preparation of this work.

Authors

Thomas E. Lodge, Ph.D., was born in Cleveland, Ohio. He received his B.A. degree with a major in zoology from Ohio Wesleyan University (1966) and his Ph.D. in biology from the University of Miami in Florida (1974). Nurturing his childhood interest in aquatic biology, he worked part-time for the Cleveland Museum of Natural History during his high school and college years. There, he became knowledgeable about the fishes of northern Ohio and contributed extensively to the museum's ichthyological collection and to its summer educational programs.

During graduate school, Dr. Lodge became fascinated with the Everglades, both academically and personally. In addition to publishing magazine articles on the Everglades, he wrote and directed an educational film ("The Everglades Region, An Ecological Study," John Wiley and Sons, 1973) and published on the fishes of the region.

After receiving his Ph.D., Dr. Lodge became an environmental consultant, specializing in wetlands and aquatic systems. In 1989, he joined Law Companies, where he is a Principal Environmental Scientist.

Dr. Lodge has led numerous environmental projects directly relating to the Everglades, including the development of methodology for evaluating the ecological function of historic Everglades wetlands. His personal interest in the region outweighs his professional activities. For more than 25 years, he has been a regular observer and photographer of Everglades wildlife.

Marjory Stoneman Douglas is the single name that has become permanently linked to the Everglades. She holds a B.A. degree from Wellesley College (1912) and seven honorary doctorates, including Litt. D. from the University of Miami and LL.D. from Barry University.

Following her arrival in Florida in 1915, she became an important force in creating Everglades National Park. Since that time, her authority in conservation—wilderness, wildlife, and water alike—has grown continuously stronger.

Her petite, five-foot-one-inch frame abruptly contrasts with her stature in the environmental arena, where she is internationally recognized as a giant by friends and adversaries alike.

A writer for all of her professional life, her best-known book is *The Everglades: River of Grass*, which was originally published in 1947 and is now in its third revised edition (1988). Her other books include *Florida: The Long Frontier, Alligator Crossing, Hurricane, Freedom River, Adventures in a Green World—The Story of David Fairchild and Barbour Lathrop, Nine Florida Stories by Marjory Stoneman Douglas,* and her recent autobiography, *Marjory Stoneman Douglas: Voice of The River.*

Preface

This project was originally conceived in 1982 as a large-format book of color photographs on the Everglades. The intent of the text was not only to accompany the excellent wildlife and habitat photographs of the Everglades by Robert Hamer, with whom I had conspired since 1971, but also to provide a more comprehensive coverage of the Everglades than available in other "coffee-table" publications. Bob and I both enjoyed "escaping" to the Everglades for photography and wildlife observation, and the project initially was much more play than work.

A first draft of the manuscript was completed in October 1988, together with a "dummy" of the book. Numerous publishers were approached, but economic considerations relating to the large format, the number of photographs, the lengthy text for that type of book, and the relatively unknown photographer and senior author met with an apologetically negative reception. The intervening years allowed for many revisions of the manuscript. Finally, at the Everglades Coalition meetings in Tallahassee during February 1993, I took a suggestion to publish the text separately (requiring extensive revision), as a service to the environmental community in its understanding of the Everglades ecosystem. However, the larger photographic work has not been forgotten.

The Text

The approach taken in this text grew out of my 27 years of experience as a biologist in southern Florida. First as a graduate student and then as a professional environmental consultant, I have been close to the Everglades, both physically and in personal and professional interests. I have developed the text around the central question, "What would one need to know about the Everglades and related ecosystems in order to have a good understanding of what they are and how they work?" Perhaps an objective would be to stand before a county commission, the state legislature, or even Congress and not be discredited for lack of knowledge. Relevant issues may include water management decisions, permits and mitigation for projects located in wetlands, water quality problems in urban and agricultural planning, wildlife management, and foremost, Everglades restoration.

The organization of the text is intended to provide useful information without forcing the reader to search too far. The order of presentation is designed to provide an orderly progression of understanding of the Everglades. However, the closing chapter, "Man and the Everglades," is written so that it can be read independently, with the hopes that its urgent messages concerning the integrity of the Everglades ecosystem might be read by people in a position to make relevant decisions concerning its fate.

References and Citations

As an aid to readers who may want to research specific topics more thoroughly, citations are indicated throughout the text and a reference list is provided at the end of the book. The intent is not to provide a reference for every statement or fact, as in a technical paper, but rather to act as a guide to appropriate sources. References that apply to entire topics are indicated with the chapter or section headings. More specific citations are indicated in the text. It should be noted that a major technical work (*Everglades: The Ecosystem and Its Restoration*, edited by Steven M. Davis and John C. Ogden, St. Lucie Press, 1994) was in preparation concurrently with this book. Following its release in January 1994, the final chapter of this handbook was largely rewritten, and citations and minor revisions were incorporated throughout the text to reflect the vast information and guidance of the restoration book.

About the Introduction

Out of professional honesty, I must admit that I "wrote" the introduction. That is, my hand put the words on paper—words dictated by Marjory Stoneman Douglas. Our agreement included my assisting her in order to accommodate her failed eyesight and necessitated reading the entire text aloud to her prior to embarking on the introduction. It was the summer of 1989 when she was "only" 99 years old. We spent numerous weekends working in her combination living room and office. During one particularly long session of my reading, I noticed her tap her "speaking" clock. It announced the time, and upon my next pause she interrupted, "I usually have a drink at 5:00. It's 4:30 now, and that's close enough." So we put papers aside and drank some Scotch whiskey.

Her attentiveness during my reading was extraordinary. Some sessions lasted almost three hours, with short breaks only for suggested changes or incidental comments. She appeared to show nothing but fascination with an admittedly lengthy text, and initially I wondered how much she had retained. Then, upon beginning the third reading session she said, "I don't believe you mentioned the harlequin snake (her vernacular for either of the Florida scarlet snake or the scarlet kingsnake) in your reptile chapter last week. I once saw one in my woodpile. I want you to include it." And so it was done, and my question of her comprehension was rested. Actually, she suggested several additions to the text, always wanting more details to satisfy her own curiosity. For whatever apprehension I had at the outset—whether we would have unpleasant disagreements, difficulty in communicating, and so forth—I now look back at those weekends with confidence: there was nothing less than friendship and lots of fun.

In the years following the initial writing, I visited Marjory numerous times. Among other things, I asked her about the statement of her age in the first sentence, as it was getting out of date. With her concurrence, the first change was merely from 99 to 100, with a sense of pride for both of us. But as time advanced without publication, she joked that I should skip directly to 102—which I did! Our final meeting to discuss the book was on March 21, 1993, just a few weeks prior to her 103rd birthday. Following that meeting, however, I realized that there was really no need to obscure those enjoyable days when we actually worked over the details of the text, and I elected to return to the original "99 years."

Thomas E. Lodge, Ph.D.
April 1994

Introduction

Marjory Stoneman Douglas

As many of you may know, I have devoted the greater portion of my 99 years to the Everglades and its related issues, particularly those having to do with that vital ingredient: *water.* Although I have been almost totally blind for some years now, I can still see clearly that the Everglades continues to need help—probably now more than ever before.

The story of my love for Florida and my concern for the Everglades begins with my father, Frank Bryant Stoneman, who had lived in Florida since 1896. First settling in Orlando, he studied for the bar and became an attorney there. But soon he became interested in the new city of Miami, where he moved in 1906—with a flatbed press he had taken for a bad debt—and started the first morning paper in Miami, which he called *The News-Record.* Without adequate money, the paper was about to fail, when Frank B. Shutts, of the law firm of Shutts and Bowen of Miami, bought control. In 1910, it was renamed *The Miami Herald,* which became the most important newspaper between North and South America.

My father started the paper at the time when Napoleon Bonaparte Broward had run for governor, under the then-popular slogan of draining the Everglades, primarily to provide new lands for agriculture. Having become aware early of water problems in the western United States, my father wrote vehement editorials protesting the idea of drainage, of the consequences of which, he said, people were entirely ignorant. This attitude so enraged Governor Broward that later, when my father won an election for circuit judge on the east coast, Broward refused to confirm his election. My father was always very humorous about that. He said it saved him from a life of shame—of having to run for election every two years thereafter.

I became interested in the Everglades immediately following my arrival in

Miami in 1915. My parents had been separated in the North when my mother, who was ill, had taken me home to my grandparents in Massachusetts, where I grew up and where I graduated from public schools and from Wellesley College. After my mother's death and my own brief unworkable marriage, I came to Florida to be with my father and to get a divorce. He had recently remarried a lovely woman who became my first friend as I settled in to establish residence with a job on the *Herald.* I found not only that the new relationship with my father would be one of the most important in my life, but that newspaper writing was the thing that I most wanted to do in the world, and that Florida, with its wonderful climate and new associations, was the place I wanted to live the rest of my life.

The city of Miami itself then contained 5000 people and was not impressive, but it was the country—the flat tropic land, the sea, the great sky, and all the excitement of a new world—that stirred my enthusiastic loyalty. I can remember that in those days, the Tamiami Trail went from Miami westward only to the Dade County line, some 40 miles. It was great fun to go out on the Trail to its end for picnics and see all beyond the wonderful untouched expanse of Everglades.

The idea that the end of the Florida peninsula should be established as a great national park had already been conceived by the man who gave the rest of his life to that pursuit. He was Ernest F. Coe, who had been a nationally known landscape architect but had been stranded penniless in Miami after the devastating 1926 hurricane interrupted the development of those early days. My father, as editor of the *Herald,* was a strong promoter of Mr. Coe's idea for a park, so that—together with Mr. Coe and others—I was immediately made a member of the first committee that worked for it. The committee was chaired by Dr. David Fairchild, and we received great support and assistance from Horace Albright, then director of the National Park Service.

The policy of the federal government was that it did not buy land for national parks. All parks had to be reserved from the public domain or donated to the federal government by states or private interests. We were 20 years working to acquire the land, originally by donation of the tract owned by the Florida Federation of Women's Clubs, but also of state lands and privately owned lands. The park was finally dedicated in 1947 by President Truman in the town of Everglades on the southwest coast. It has become the second most visited national park in the country. I am, at present, the only living member of the original committee.

We know now that the basin of the Everglades begins with the once-meandering flow of the Kissimmee River down into Lake Okeechobee from the north. Prior to man-made changes of the lake's drainage, Lake Okeechobee overflowed its southern rim during the wet season into the broad expanse of the Everglades. It all worked as a unit: the Kissimmee–Lake Okeechobee–Everglades watershed. But beginning in 1882, this unit was changed. First, the Caloosahatchee River was "canalized" and connected to the lake, making an open outlet for the lake's waters toward the west to the Gulf of Mexico. Then, great canals were excavated right through the Everglades, further diverting would-be Everglades waters from the lake southeastward to the Atlantic coast. These were the Miami, the

North New River, the Hillsborough, and the West Palm Beach canals. In 1916, a final canal—the St. Lucie—was begun, adding a more direct, eastward outlet from Okeechobee to the Atlantic by way of the Indian River at the town of Stuart. The lake's entire southern rim was diked by a huge levee, so that the only outlets were the canals, all fitted with gates to control the waters in an effort to put man—not nature—in charge of the Everglades. All this provided an enormous area of reclaimed land upon which agriculture, mostly sugarcane, developed to the south and southeast of the lake.

Through the years, the canals have been enlarged and interconnected into a vast network, and even the once-meandering beauty of the Kissimmee and Caloosahatchee has been lost to dredges and straight-line engineering. Today, the original, natural order of flow has been so interrupted by canals, levees, and water-control structures (dams, gates, locks, and even pumps) that the movement of water through the watershed towards Everglades National Park has become entirely unnatural. Nevertheless, the condition of the park still depends upon the proper maintenance of the Kissimmee–Lake Okeechobee–Everglades watershed. It is ironic that we have now started to undo past mistakes on the Kissimmee River by reconnecting some of its original meanders and allowing its waters again to spread out over parts of its once natural, pollutant-cleansing flood plain.

With a growing number of others, my voice has been heard through many decades in the efforts to preserve the historic flows of water to the park. Today, the powerful Washington-based *Everglades Coalition*—comprised of over 20 national, state, and local organizations, including my own *Friends of the Everglades*—continues this work to strengthen the control of water for the Everglades and to the park. But the coalition's work must not be taken for granted. Its continuing success requires a never-ending fight against the forces of agricultural and commercial development. The future of this wonderful park depends on the public support of these conservation efforts. In that regard, I hope this book will prove to be an important contribution to the understanding of the nature and beauty of the Everglades as well as to the many-faceted challenges of keeping the Everglades alive and well.

Part I
Background

1

The Everglades in Space and Time

I t could be called an accident of geography that the Florida peninsula extends from North America's temperate climate to the edge of the Caribbean tropics (Figure 1.1). South (or southern) Florida—the portion of the peninsula from Lake Okeechobee southward—is the home of the Everglades and is especially appreciated because of its proximity to the tropics. Delighted by its warm weather and tropical vegetation, few people would question the geologic origin of the region. Among the few are geologists studying the region's "basement" rocks, lying as deep as three miles beneath the surface. Those ancient rocks have an intriguing story to tell about continental drift or, more properly, plate tectonics.

Florida, Geologic Time, and Plate Tectonics[150, 157, 162, 214, 259]

When the earth's first fishes were evolving, the predecessors of the modern continents of North America and Africa were in positions comparable to those of today. Africa and South America, on the other hand, were connected as parts of a larger continent called Gondwana by geologists, joined below what is now Africa's west bulge. (Gondwana also included Antarctica, Australia, and India, which adjoined the east side of Africa.) Between North America and Gondwana was an ocean, known to geologists as the Iapetus, in a position similar to that of the modern North Atlantic. A map of ancient North America would have lacked a prominent modern feature: Florida. The terrain that was to become Florida was then part of the west bulge of Africa.

In the millions of years that followed, North America slowly approached Gondwana, narrowing the Iapetus Ocean until the land masses collided about

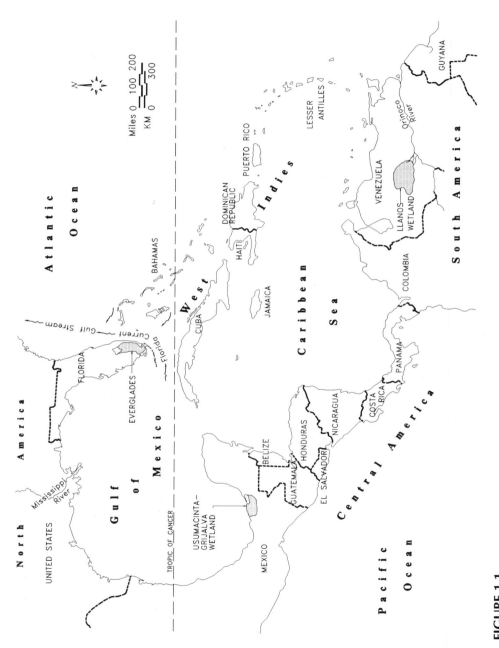

FIGURE 1.1
The Everglades in the Gulf of Mexico and Caribbean regions.

300 million years ago. The collision created the Appalachian Mountains as it fused North America's east coast with Africa's west bulge, just north of South America's position,* and locked Florida's terrain deep in the interior of the giant continent that was formed. Geologists call that supercontinent Pangaea.

After a long period of quiet, about 200 million years ago while early reptiles were evolving (the Triassic period), Pangaea began to tremble with volcanic activity that centered near Florida. Volcanic eruptions added massive igneous rock, transforming Florida's earlier sedimentary terrain. The upheaval continued, and about 180 million years ago (the Jurassic period) it began forcing North America to rift from Pangaea, giving violent birth to the North Atlantic Ocean, the Caribbean Sea, and the Gulf of Mexico. In the new configuration, however, the shallow marine terrain of Florida, the Bahamas, and parts of southern Georgia and Alabama was welded to the North American continent.[113]

The shallow marine environment of the youthful Florida platform produced sediments rapidly. The sediments later formed thick sequences of limestones and related carbonate rocks. One sequence, the Sunniland Formation, is of particular interest. It is about 135 million years old (lower Cretaceous time) and now lies 11,000 feet beneath the Big Cypress Swamp and western Everglades.** It is the modest oil-bearing formation of the region. Production comes from wells located along the northwest-southeast, reef-like Sunniland Trend, a series of carbonate mounds primarily composed of ancient oyster-like mollusks called rudists.[66, 130]

Upheaval in the remainder of Pangaea continued, and about 125 million years ago, South America and Africa began to separate, opening the South Atlantic and completing the initial formation of the Atlantic Ocean. Opposing motions of the earth's crust between North and South America created the West Indies in the evolving Caribbean, as the two continents continued their departure from Africa while the Atlantic continued to grow. Unlike its configuration today, the Caribbean Sea was open to the Pacific as well as the Atlantic Ocean. Only in the last 2.5 million years have geologic movements uplifted the area of Panama and Costa Rica, completing the land bridge through Central America that now connects North and South America and separates the Pacific from the Caribbean. Even today, the earth's crust continues to move. Volcanos in Central America are evidence of continental uplift, and the Atlantic Ocean grows an inch or so wider every year.[182, 256]

As a result of its long submerged history, the Florida peninsula is now a broad platform (the Floridan Plateau) built of stable sedimentary rocks (principally

---

* The location of the actual collision, which occurred during the Paleozoic era, is termed the Alleghanian suture and is thought to be represented by an area of magnetic deviation called the Brunswick magnetic anomaly. This feature runs from offshore along the southeastern coast of the United States, through southern Georgia and southeastern Alabama to the western panhandle of Florida.[226]

** With abundant evidence that the sedimentary accumulations on the Florida platform developed in shallow water, it is apparent that the basement itself has been subsiding through most of its history. The accumulation of sediments has merely kept pace with the rate of subsidence.[157]

limestones ranging from ancient to very recent age), layered over the ancient basement of African origin. To the east, the plateau drops off abruptly into the Atlantic. Southward, it slopes gently to a "rim" occupied by the Florida Keys and then also drops off quickly into the deep trough known as the Straits of Florida, which is located between Cuba and Florida and carries the Florida Current.* To the west, the plateau slopes gradually far out into the Gulf of Mexico before receding into deep water. South of Lake Okeechobee, this plateau is so flat that only the direction of water flow can indicate which way is downhill.[100]

Land, Sea Level, and Climate[100, 138, 185, 256]

It is obvious that the present configuration of the Florida peninsula—the land area depicted on any conventional map—has changed considerably through geologic history. Sedimentary rocks throughout the peninsula reveal that it has been submerged as a shallow platform for much longer than it has been emerged. The sporadic terrestrial history of the region began only 25 million years ago (the Miocene epoch), as indicated by fossils of terrestrial animals. Land building was assisted by sediments of terrestrial origin (clastics), especially silica sand from the Appalachian Mountains. These materials were transported down rivers and then southward over the peninsula by currents, waves, and wind. The process endowed the northern and central portions of the peninsula with beach ridges and dunes, thus creating the present undulating topography. Less of the materials reached southern Florida, however, leaving it with the lowest and flattest elevations and therefore the least time with its surface above the sea.[163]

The emergence of land in the Florida peninsula was aided by the accumulation of sediments, but the most impressive factor was cycles of glaciation in the earth's high latitudes. During the last two million years (the glacial age or Quaternary period) several episodes of glaciation occurred,** when the Florida peninsula stood much higher above sea level than it does today and the climate was cool and dry. Between the glacial cycles, marine waters inundated the region, sometimes to depths exceeding 100 feet,[185] and the climate was warm and humid.

---

* Most people, including Florida fishermen, refer to this current as the Gulf Stream, or simply "the Stream." To remain technically correct, the Gulf Stream should be referred to as the great northward-flowing current of the North Atlantic, formed by the confluence of the Florida Current and a westwardly current that arrives off the central Florida coast, mostly north of the Bahamas.[106a, 108]

** Continental drift is thought to have been responsible for the rather recent configuration of continents and oceans that has restricted circulation of currents to and from the Arctic Ocean. This condition, coupled with the position of Antarctica squarely over the South Pole, has probably caused the glacial age that began about two million years ago. The glaciations (actual accumulations and advances of polar ice) during this age are thought to result from cyclic interactions (Milankovitch cycles) of changes in the earth's elliptical orbit and the tilt of its axis.[184]

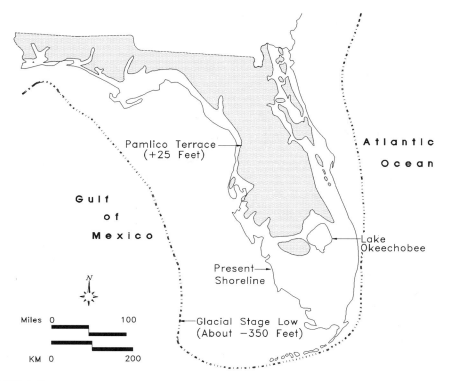

FIGURE 1.2
Two former shorelines of Florida and the present shoreline (the Pamlico Terrace, about
125,000 years ago), and the last glacial stage low (about 20,000 years ago). (Adapted from
Hoffmeister[100] and NOAA.[166])

Step back in time to the beginning of the last glacial age, about 100,000 years
ago, which is relatively recent history for the Floridan Plateau. To put that time
increment—100,000 years—in perspective, about 700 of those increments have
passed since the dinosaurs perished, and 1800 have passed since Florida's basement terrain was forced from Africa. Sea level just over 100,000 years ago was 25
feet higher than it is today. It stood at that level for a long time during the last
warm interglacial period, and the final sediments that form the surface of the
Everglades region were deposited, including the Anastasia Formation (the northeastern Everglades), the Fort Thompson Formation (the northern and central
Everglades), the Miami Limestone (the central and southern Everglades, Florida
Bay, and the Lower Florida Keys), and the Key Largo Limestone (the middle and
upper Florida Keys). Waves then carved the well-known Pamlico Terrace, a
modern topographic feature representing the shoreline of that time. This terrace
is recognizable around much of the peninsula at the +25-foot contour (Figure 1.2).
The southernmost land was near the town of Immokalee, more than 100 miles
north of Key West. During earlier interglacial times, sea level was even higher
and the peninsula only half of its present length.[157]

The onset of the most recent glaciation drastically reversed Florida's character. At the glacial maximum only 17,000 years ago, the north polar ice cap reached as far south as New York City, the location of Chicago lay under a mile of ice, and the basins of the Great Lakes were gouged out by the southward motion of mountainous ice. So much water was removed from the world's oceans—locked up in ice and snow piled above sea level in the ice caps of the northern and southern hemispheres—that sea level was between 300 and 400 feet lower than it is today. The lower sea level exposed much more of the Floridan Plateau, which encroached far into the Gulf of Mexico. The peninsula was about twice as wide as it is today, but it did not extend significantly eastward or much farther south than the Florida Keys, because it was constrained by the deep Straits of Florida. Based on several lines of evidence, southern Florida was never connected to islands of the Caribbean region.[182, 251]

The periods of glaciation also had important climatic effects. The Florida peninsula was substantially cooler and drier than it is today, and surface waters (lakes and streams) were probably rare or absent because of the leaky characteristic of the limestone. The climate was not tropical, and plants and animals that required tropical conditions were unable to persist, even in the southernmost parts. Alternatively, fossil evidence shows that animals, such as the porcupine, and numerous plants that are now found only in areas far north of Florida occurred in the Florida peninsula at that time.[163, 200]

The warming trend that followed the last glacial cycle again changed Florida. Sea level rose rapidly, shrinking the area of land. About 6000 years ago tropical plants again began to flourish in extreme southern Florida, invading from the West Indies as they had in past interglacial times. Rainfall increased, and the Everglades finally came into existence. The story of the Everglades is but a minuscule chapter—the last 5000 years—in the vast realm of time glimpsed in the origin of Florida.

2

An Ecosystem Overview

arjory Stoneman Douglas described the Everglades as the "River of Grass," and before her, the Miccosukee Indians called it "Pa hay okee," meaning grassy water. Why have special names been given to and entire books written about the Everglades?

The Everglades is unique to Florida, and the term itself requires some comment. Technically it refers to the expanses of freshwater marsh originally extending from Lake Okeechobee to nearly the southern tip of the Florida mainland. The word *Everglades* has an obscure and apparently accidental origin, with the first part, *ever*, originally indicating river. The second part, *glade*, is probably the English word meaning an opening in a forest where grasses cover the ground.

Marjory Stoneman Douglas traced these origins in her famous book, *The Everglades: River of Grass*, published in 1947 (its 1988 revised edition is still in print). She identified the first published appearance of the term as an 1823 map. Since our introduction at The Kampong* in Coconut Grove in 1972, I have appreciated Marjory's intolerance of improperly researched work. I have therefore had to consider carefully my use of the term "Everglades." Marjory's book begins, "There are no other Everglades in the world." Because she treats the word as plural, I have had to choose between singular and plural, as it is found both ways in the Everglades literature. My decision is to treat *Everglades* as singular because it is the name of one physiographic region; I can, however, agree with

---

* The former estate of Dr. David Fairchild (1869–1954), botanist and plant explorer for whom the famous Fairchild Tropical Garden, located near Miami, was named. In 1963, after the death of Mrs. Fairchild, Catherine H. Sweeney purchased and maintained The Kampong in the spirit of Dr. Fairchild's interests. In 1984, she donated it to the Pacific Tropical Botanical Garden, which was then renamed the National Tropical Botanical Garden by act of Congress specifically for the acquisition of The Kampong, to be maintained as a tropical garden in perpetuity. Over the years, a great many people involved in botanical and related work have met at The Kampong, and some of the earliest meetings to lay the framework for creating a national park in the Everglades were held there.[55, 209, 223]

Marjory that the Everglades is truly unique! (Actually, she now agrees that singular is appropriate.)

This book focuses on the Everglades in terms of its relatively recent origin and its probable future. Considerable attention is devoted to the types of environments that surround it, which are inextricably linked biologically, geologically, and hydrologically. The Everglades must be understood within the context of a larger region, which encompasses more than just southern Florida. The historic watershed of the Everglades came from as far north as the Orlando area of central Florida.

A Unique and Valuable Ecosystem

The state of Florida contains enormous acreages of marshlands, ranging in size from tiny isolated depressions within pinelands to the vastness of the Everglades, which was originally about 4000 square miles in size. The Everglades is not special just to Florida—it is unique in the world. Most of the world's larger wetlands receive their water and nutrients from associated rivers that overflow their banks. Examples that superficially resemble the Everglades (some even much larger in size) are the Pantanal of Brazil, the Llanos of Venezuela,[174] and especially the delta region of the Usumacinta and Grijalva rivers in southern Mexico.[30, 170] Except for the historic overflows from Lake Okeechobee, which principally affected the northern Everglades, the bulk of the Everglades ecosystem received nutrients only from the atmosphere, primarily in the form of rainfall. Thus, the configuration of the Everglades as a "sheet flow" ecosystem, independent of river or stream channels, is one of a kind.[92]

The Everglades would probably arouse little interest if not for the values associated with it. These values include the enormous populations of water birds (Figure 2.1) (which is the main reason that an earlier generation sought to protect the area as a national park), an abundant, high-quality water supply for the human population of Florida's southeastern coast, and agriculture. How the Everglades works will be described in the following chapters. In the final chapter, in which it is made obvious that these values have not been in harmony, a pathway for restoration of the damaged ecosystem is recommended.

Terms and Definitions

Some important definitions are appropriate at this point. First is the term *wetland*, which generally describes areas of the earth's surface that are regularly or periodically covered by shallow water. Legal definitions may include lands where little or no surface water occurs, but where ground water periodically rises close enough to the surface to saturate the soil, making a *hydric soil*, which favors a prevalence of wetland plants and restricts the growth of upland plants.

Most wetlands alternate between flooded and "dry" (i.e., lacking surface

FIGURE 2.1
The essence of the Everglades: a mixed group of wading birds feeds just north of the Tamiami Trail, in Water Conservation Area 3A, February 1973. (Photo by R. Hamer.)

water) on a regular basis, as with tidal cycles or seasonal cycles of changing rainfall. For non-tidal wetlands, the average annual duration of flooding is called the *hydroperiod*, which is based only on the presence of surface water and not its depth. Relative to the Everglades, *long hydroperiod* is in excess of ten months (often with continuous flooding for a few years), and *short hydroperiod* is about seven or fewer months, but large annual variations are typical of individual locations because of year-to-year differences in rainfall.

A less frequently used but nonetheless important term that refers to depth as well as hydroperiod is *hydropattern*.[91] Hydropatterns are best understood by a graphic depiction of water level (above as well as below the ground) through annual cycles. Typical hydropatterns for the Everglades show water one to one-and-a-half feet deep for several months during the wet season and a foot or so below the ground at the end of the dry season. The various areas differ greatly, however, and annual and multi-year cycles are highly variable.

Wetlands come in many varieties, but the two main types are *swamps* and *marshes*. Although the definitions are not universal, swamps are usually considered to be wetlands in which the dominant plants are trees. Examples of swamps in the Everglades region include the mangrove swamps in the tidal brackish and saltwater areas and the cypress swamps in freshwater areas. If trees are so scattered that their shade is insignificant, the area is better called a marsh. Marshes are dominated by much lower, herbaceous (non-woody) vegetation and often look like prairies.

Strictly speaking, the term *Everglades* applies only to southern Florida's huge, interior freshwater marsh variously dotted with "islands" of trees. The term is often shortened to just *the glades*. Everglades National Park is situated at the southern end of the Everglades, and it encompasses many more types of habitats than just "Everglades." In fact, the true Everglades occupies only a small portion of the park. Much larger portions are comprised of the marine waters of Florida Bay and the mangrove-dominated estuaries of the southwest tip of the Florida mainland. The principal kinds of environments of the Everglades region* are

- Freshwater marshes
- Wetland tree islands (broad-leaved types)
- Cypress heads, domes, and dwarf cypress forests
- Tropical hardwood hammocks
- Pinelands
- Mangrove swamps and mangrove islands
- Coastal saline flats, prairies, and forests
- Tidal creeks and bays
- Shallow, coastal marine waters

Other authors may use various different names for these environments or may even use other kinds of divisions.[39, 92, 140, 175] It is important to bear in mind that any system of categorizing life's realities is only for convenience. We often turn the situation around, wanting nature to conform to our system of nomenclature, only to become frustrated with its lack of cooperation. Not everything fits into our preconceived notions of how nature "should" be organized. Thus, while these divisions are useful for organizing our thoughts, there will be some exceptions: places that do not conform to these divisions and divisions that might better be subdivided for particular purposes.

The Historic Everglades

As described in the introduction, before changes began in 1882, the entire ecosystem that included the Everglades was a watershed, beginning near the present location of Orlando: the Kissimmee–Lake Okeechobee–Everglades watershed (Figure 2.2) (often given the acronym KLOE). Surface water flow from

---

* The historic Everglades also included a pond apple swamp around the south and southeast shore of Lake Okeechobee and a narrow band of cypress swamp along the eastern edge primarily in Palm Beach and Broward counties.[46, 47, 92] These communities no longer exist (see the following section on the historic Everglades and Chapter 4) and are not listed here.

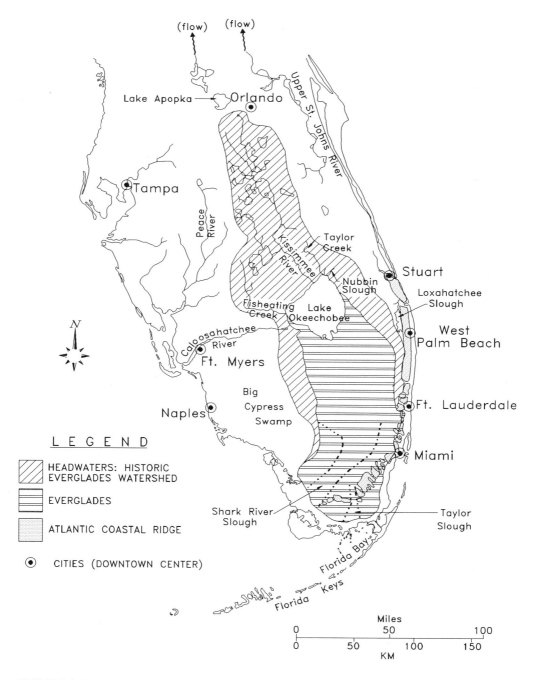

FIGURE 2.2
Map of the historic Everglades watershed.

the Kissimmee River, together with flows from Fisheating Creek, Taylor Creek, and Nubbin Slough, ended in Lake Okeechobee, from which there was no outlet during the dry season. In normal summer rainy seasons, however, the water level of Lake Okeechobee rose; upon reaching an elevation of at least 15 feet (above sea level), it began overflowing the southern rim of the lake through a forest of pond apple trees (also called custard apple [*Annona glabra*]). Attempts by early explorers to find navigable outlets from the lake were frustrated by this swamp, fringing the south and southeastern rim of the lake like a necklace up to three miles thick. All streams leading from the lake ended in the profusion of swamp forest. Beyond the pond apple swamp* lay the sawgrass-dominated Everglades, some 40 miles wide and extending southward about 100 miles to the tidal waters of Florida Bay and the Gulf of Mexico.**, [21, 46, 177, 229]

Sawgrass was far from the only plant in this vast ecosystem. It completely dominated a zone that extended about 15 miles south and southeast from the custard apple swamp before giving way to a more varied landscape that included open-water sloughs (pronounced "slews") where lily pads adorned the surface, tree islands, islands of sawgrass, and other marsh communities. This mosaic covered the major portion of the Everglades ecosystem (Figure 2.3).

Although some marsh communities were far easier to traverse on foot or in canoes, there were no open channels through the Everglades. The water moved in a vast "sheet." The slope from Lake Okeechobee to tidal waters averaged less than two inches per mile. With slightly higher terrain in the Big Cypress Swamp to the west and the Atlantic Coastal Ridge to the east (highest elevations scarcely over 20 feet), the Everglades waters were guided slowly southward by the imperceptible slope of the terrain.[39, 177] The rate of flow, slowed by the marsh vegetation, ranged from negligible to about two feet per minute.[204, 248] The depth varied considerably with seasons and years, but rarely exceeded four feet in the deepest marshes. Average wet-season maximum depths were probably between one and two feet.***, [90]

---

* The first complete map of the vegetation of the Everglades, published in 1943,[46] also showed a willow and elderberry zone between the pond apple swamp and sawgrass. Not noted by the early explorers, this zone may have resulted from the initial drainage, and attendant fires, in the northern Everglades.

** It is noteworthy that passage through the swamp and into the Everglades was possible in canoes during very high water and that the Everglades itself was regularly traversed by the Indians (and a few early explorers) in such craft except in very low water.[228, 229, 258]

*** Flooding in 1947 rendered much of the Everglades six to eight feet deep.[177] Water depths in the original Everglades will never be known, but based on descriptions of the vegetation, such excessive depths must have been unusual and of short duration. The depths given here are from an evaluation of 32 years (1954–85) of data obtained in Everglades National Park. Because these data were collected after the Everglades was subject to water management control, they are undoubtedly biased to some extent by control of flooding extremes and by water retention north (upstream) of the park during drought conditions. Nevertheless, the recorded levels correlate with the general integrity of most wetland plant communities but not the animal communities, as discussed in the final chapter.

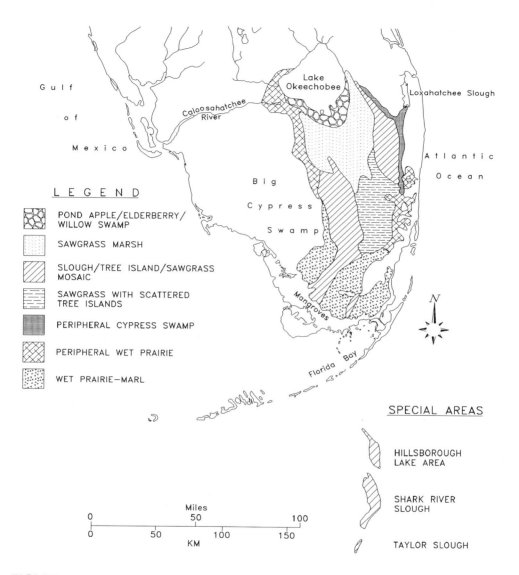

FIGURE 2.3
Map of the major plant communities of the historic Everglades. Note that the western boundary of the Everglades, against the Big Cypress Swamp and lands farther north, is indefinite: it merges into other wetlands. (Modified from Davis.[46])

The shallow water also meant that small variations in local topography were important: differences of a few inches in elevation caused large responses in vegetation types.[39] Ranging from the subtle mosaic of freshwater marsh and tree island communities, to low pinelands along the Atlantic Coastal Ridge, and finally to the saline environments of the coast, these variations added intriguing dimensions to the landscapes of the region. The interesting details of the various plant communities that comprise the subsystems of the Everglades region are discussed in subsequent chapters.

Origin of the Everglades

The environmental factors that controlled the ecology of the historic Everglades—the abundance and seasonality of rainfall, water quality, fire, and unique geographic/geologic features—were also responsible for its evolution from the uplands that existed when the sea level was much lower. This evolution began very recently in geologic time, with the accumulation of peat soil in the Everglades starting only about 5000 years ago.[84] From its much lower stand during the glacial period, sea level had risen enough so that the area that was to become the Everglades could no longer drain rapidly; Florida's rainfall had increased to the point where it resembled today's, a much wetter climate than had existed earlier. The presence of surface water began to stress then-existing upland plants and to favor the growth of marsh plants. Under the protection of water, wetland soils began to accumulate: the Everglades was "born."

A region dominated by upland plant communities (including pinelands perhaps like those existing today along the eastern edges of the Everglades) began to disappear.[138, 175] First in the areas of lowest elevation and then progressively higher, the Everglades vegetation advanced; its developing soils covered the original scanty soil and widely exposed limestone bedrock, the "floor" of today's Everglades. Because the accumulation of wetland soils depended upon water level, the developing terrain was very flat, far more level than the original rock floor beneath.[46]

This "growth" of the entire Everglades proceeded to the level where wet season waters began to spill over into the lower areas of the Atlantic Coastal Ridge. Some water passed from the northern Everglades eastward through the Loxahatchee Slough (in Palm Beach County) to the Atlantic. Farther south, overflow produced some short rivers, such as the New River in Fort Lauderdale and the Miami River; in Broward and Dade counties, numerous "transverse glades" (wet season overflows resembling Everglades habitat) were produced. A transverse glade existed, for example, near the present location of the South Miami Hospital, where high water drained from the Everglades to Biscayne Bay.[94]

Thus, given the elevation of the Atlantic Coastal Ridge, the Everglades evolved to its maximum possible extent, the giant wetland that existed when the first Europeans visited Florida. Native Americans, however, saw it all. They inhabited southern Florida for at least 5000 years before the first soils of the Everglades were formed.[28]

Part II

Environments of the Everglades Region

3

Freshwater Marshes: Water, Weather, and Fire

Expanses of freshwater marsh are the essence of the Everglades. The productivity of these marshlands, together with the habitat they provide, comprise the basis of the value of the Everglades to wildlife. While the expanses of marshes often give the impression of tranquility, nature has maintained them through powerful stresses imposed by water, weather, and fire.

Marsh Vegetation and Plant Communities[39, 46, 92, 140, 175]

If a single word had to be used to describe the Everglades, it would be *sawgrass*, which is not a true grass but rather a member of the closely related sedge family (Figure 3.1). It is named for the sharp, upward-pointing teeth that line each edge of the leaf's stiff "V" cross section. Once its highly perishable seedlings become established, sawgrass is a very tough plant, well adapted to the rigors of the Everglades. Its principal enemies are soil fire during drought, multi-year flooding deeper than about a foot, and fire during rising water. The latter condition eliminates the oxygen supply for its roots, causing suffocation. Essentially nothing eats sawgrass, although the abundant, late summer seeds are eaten by ducks. Most sawgrass reproduction is by lateral spread of its prolific roots.[3, 5, 70, 261]

Many marsh plants in addition to sawgrass occur in the Everglades (Table 3.1), with over 100 species present. From a biogeographic perspective, most of these species, including sawgrass, have such extensive ranges in temperate and tropical regions that the distinction is unimportant. Overall, however, the marsh flora is more related to temperate North America, with but a few examples (such as flag) of tropical origin.[139]

FIGURE 3.1
Sawgrass in bloom. Typically, five or six of these flower clusters are arranged in a line along the flowering stem. (Photo by R. Hamer.)

TABLE 3.1
Common or Conspicuous Freshwater Marsh Plants of the Everglades

Common name[a]	Scientific name[b]
arrowhead	*Sagittaria lancifolia*
common cattail	*Typha latifolia*
muhly grass (or hairgrass)	*Muhlenbergia capillaris*
maidencane	*Panicum hemitomon*
spike rush	*Eleocharis cellulosa*
white-top sedge	*Dichromena colorata*
sawgrass	*Cladium jamaicense*
Tracy's beak rush	*Rhynchospora tracyi*
inundated beak rush	*Rhynchospora inundata*
pickerel weed	*Pontederia cordata*
spider lily	*Hymenocallis latifolia*
swamp lily	*Crinum americanum*
flag (fireflag or arrowroot)	*Thalia geniculata*
spatterdock (yellow water lily)	*Nuphar luteum*
white water lily	*Nymphaea odorata*
floating heart	*Nymphoides aquatica*
water hyssop	*Bacopa caroliniana*
bladderwort	*Utricularia foliosa*

[a] The common names follow the author's personal preference. Common names of most marsh plants are not standardized and vary among sources. In some cases (e.g., the genus *Rhynchospora*), the species are so difficult to identify that only scientific specialists deal with the names. Most specialists are professionally comfortable using only the scientific names, and thus common names have never come into regular use. On the other hand, some species are widely known by non-scientists, yet have not been standardized. A case in point is sawgrass, which is used here as one word following numerous sources such as Olmsted and Loope[175] and Robertson.[197] However, "saw grass" (Long and Lakela[139]) and "saw-grass" (Godfrey and Wooten[85]) are also embedded in the Everglades literature. This author considers the latter forms cumbersome.

[b] Scientific names follow Godfrey and Wooten.[85]

Discrete variations in marsh vegetation over the expanses of the Everglades have prompted numerous efforts to define plant communities. Regional variations result from geologic and hydrologic conditions, but many localized variations do not have obvious causes. Because of the region's pronounced wet and dry seasons, conditions promoting and restricting growth of various kinds of marsh plants are always at work: the rigors of excessive flooding and the opposite, concurrent drought and fire. These stresses have caused the mosaic pattern of growth seen throughout the Everglades, which is most easily appreciated from an aerial view. The following categories are generally recognizable and widely used (Figure 3.2):

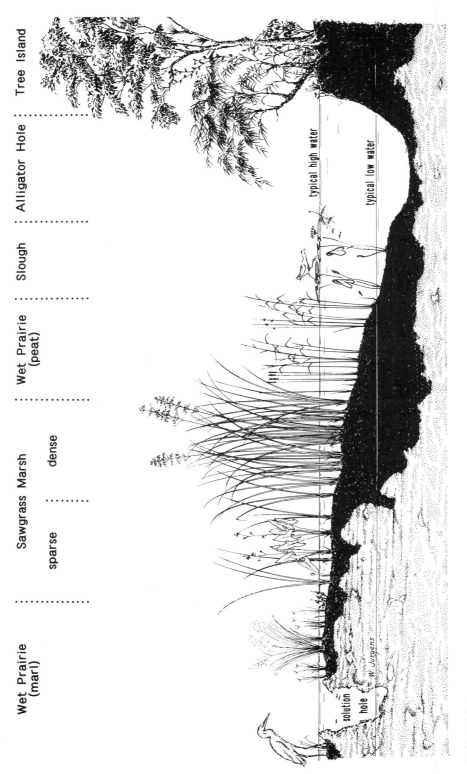

FIGURE 3.2
An idealized, greatly compressed section of typical Everglades plant communities.

- Sawgrass marsh (grading from dense to sparse)
- Wet prairies (two different kinds)
- Slough (pronounced "slew")
- Pond and creek

Each provides different and important ecological functions.

Sawgrass Marshes

Sawgrass marshes are areas overwhelmingly dominated by sawgrass, with little other conspicuous vegetation. Spacing of the plants grades from dense to sparse (Figures 3.3 and 3.4). Dense, tall sawgrass originally covered most of the northern Everglades, where it grew to nine feet tall in deep peat soils. Most of that area is now used for agriculture, mainly sugarcane, but some remains protected in conservation areas. In the central and southern Everglades, sawgrass marsh typically occurs in patches; its stature is not so tall (usually less than five feet), and several other marsh species may occur together with the sawgrass.

FIGURE 3.3
Sparse sawgrass and a mat of floating algae (periphyton) growing over bladderwort in Water Conservation Area 3B. (Photo by T. Lodge.)

FIGURE 3.4
Seen from the air, elongated patches of dense sawgrass in Water Conservation Area 3A are surrounded by sparse sawgrass where abundant periphyton has formed a floating mat. The dark trails are the work of alligators and deer that have parted the periphyton in their treks through the marsh. (Photo by T. Lodge.)

The average hydroperiod for sawgrass marsh is about ten months, but it ranges from less than six months to nearly continuous flooding, and typical wet season depths range from one to one-and-a-half feet. (Hydroperiod/hydropattern estimates for sawgrass marsh as well as the other marsh types in this chapter are based on Gunderson,[91] Kushlan,[116] and Olmsted and Loope.[175]) The deeper water and longer hydroperiod support taller, dense sawgrass, with the drier end of the range supporting more open, sparse growths. Dense sawgrass in particular harbors little animal life, but it is a habitat where alligators often build their nests.[137, 170]

Wet Prairies

Wet prairies are areas of marsh dominated by emergent plants other than sawgrass (and only superficially resembling prairie habitat of the Great Plains). The term *wet prairie* is not standardized in the Everglades literature and has been applied to two distinct communities. In common, these two communities have lower and less dense vegetation than sawgrass marsh, providing more open water for aquatic animals.[137]

One type of wet prairie, often called *marl prairie,* occurs on thin calcitic soil (marl) over limestone bedrock, which may be exposed as jagged, foot-tall projections called *pinnacle rock* or dissolved below the surface into pockets called *solution holes.* The hydroperiod is the shortest of all the marsh types in the Everglades, averaging between three and seven months. This community flanks both sides of the southern Everglades, where it was once much more extensive, especially in Dade County. It virtually always contains some low sawgrass (less than three feet tall), but plant diversity is high. About 100 species occur there, and thus the alternate term *mixed prairie.* Common species are beak rushes, spike rush, white-top sedge, and muhly grass. The latter can be recognized in summer by its pink inflorescence, which sometimes gives entire vistas a pink glow. An interesting variation of marl prairie contains scattered, dwarf cypress trees and is called *dwarf cypress prairie,* the most bizarre landscape in South Florida (Figure 3.5).

Certain marl prairies can be important short-hydroperiod feeding areas for wading birds, supplying prey early in the dry season (October into January). Crayfish and amphibians, which do not require permanent surface water, are normally common, but in areas that contain numerous solution holes (the "rocky glades"), small fishes, applesnails, and other prey may also be abundant, having survived "underground" during the dry season. Without solution holes that reach into the dry season water table, marl prairies have less ecological value. As a cautionary note, solution holes may go unseen during the wet season, covered by floating vegetation or mats of algae. For this reason, the marshes along the main Everglades National Park road between Taylor Slough and Rock Reef Pass are very hazardous (Figure 3.6).[136, 172]

The other type of wet prairie is a much deeper marsh community developed on peat soil, characteristically with a long hydroperiod and lower plant diversity.[86] Maidencane, Tracy's beak rush, or spike rush usually dominate. These areas more commonly occur on peat soils in the northern and central Everglades,

FIGURE 3.5
Dwarf cypress prairie following a dry season rainstorm near Pa-Hay-Okee Overlook, Everglades
National Park. Cypress in the Everglades and neighboring Big Cypress Swamp typically loose
their leaves in late November and develop new leaves in March, prior to the end of the dry
season. (Photo by R. Hamer.)

where they characteristically lie between sawgrass marshes and sloughs. They
are important for fish and aquatic invertebrates, such as prawns, which require
more permanent water. This habitat provides abundant prey for wading birds
toward the end of the dry season, typically March and April.

Sloughs

These deepest marsh communities are the main avenues of water flow through
the Everglades. The hydroperiod is about 11 months, but flooding may be continuous for several years. Water depth in the rainy season may be over three feet,
and the annual average is almost a foot, which is significantly deeper than the
other marsh habitats. Like the longer hydroperiod type of wet prairie, sloughs

FIGURE 3.6
A solution hole in a marl prairie community of the "rocky glades" west of Taylor Slough,
Everglades National Park. Photographed in the late dry season, this two-foot-deep hole still
contained several inches of water and several small fish. (Photo by R. Hamer.)

occur over peat soil and support an abundance of fishes and aquatic invertebrates.[135] The dominant vegetation includes submerged and floating aquatic
plants such as bladderwort, white water lily, floating heart, and spatterdock.
Normally, little or no sawgrass is found, but maidencane may be abundant
(Figures 3.7 and 3.8).

FIGURE 3.7
Summertime thunderstorm over the Everglades. Spatterdock thrives in an artificial slough at the Anhinga Trail, Everglades National Park, an area that is maintained by the Park Service and long predates the park as part of Royal Palm State Park, established in 1915. Sawgrass marsh and tree islands lie in the background, all in Taylor Slough. (Photo by R. Hamer.)

FIGURE 3.8
Seen from the air, patches of sawgrass marsh (partially invaded by willow at left) in Water
Conservation Area 3A within extensive areas of slough vegetation, dominated here by water
lilies, which appear as white pepper dots. (Photo by T. Lodge.)

Areas where slough communities are abundant are characteristically dotted
with tree islands. Tree islands (see Chapter 4) and sloughs characterize part of the
northern Everglades (now within the Arthur R. Marshall Loxahatchee National
Wildlife Refuge and historically called the Hillsborough Lake area because of the
abundance of open-water slough habitat) and the Shark River Slough of the
central and southern Everglades. The Shark River Slough is over 20 miles wide
in the south-central Everglades, narrowing to about 6 miles through Everglades
National Park. It is named for one of several tidal rivers that receives its water in
the coastal area of the park. Taylor Slough is a small slough on the east side of the
southern Everglades.

Ponds (Alligator Holes) and Creeks

Small, open-water areas are scattered through most of the Everglades (Figure 3.9). They are flooded continuously except during unusually severe droughts. They occur mostly in marshes that have a longer hydroperiod and not in marl prairies. Some may have originated as a result of a fire burning a pocket in an area of peat soil during an extended drought, and others may have been initiated by alligators in areas of marsh. Whatever the origin, alligators perpetuate these ponds by their maintenance activities. The centers of active alligator holes are relatively free of vegetation, but near the edges there may be spatterdock and other species of aquatic and floating plants similar to those of the sloughs, and wetland trees take root on the banks. Alligator holes (see the alligator section in Chapter 15) are very important as refugia* for aquatic life during the dry season. In the extreme southern Everglades, where fresh water and tidal water meet, numerous creeks (see Chapter 7 on mangrove swamps) function like alligator holes and are maintained by alligators. They perform the same important function for wildlife as alligator holes.[137]

Many species of marsh plants can be found in more than one community. Spike rush, for example, is common in wet prairies and sloughs, and it quickly invades areas where sawgrass has been killed by a soil fire or by salt water from an unusual storm tide near the coast.[2] Spike rush leaves are nearly round in cross section. Initially green, they become brownish-orange upon aging, and early morning or late afternoon sunlight can produce a stunning visual effect in spike rush areas.

The ability to recognize certain kinds of plants, such as cattails and flag, can make hiking through Everglades marshes more pleasant. Most people recognize the tall, smooth cattail leaves, but few are familiar with flag, a plant with large, light green oval leaves that may be over two feet long. The presence of these two species often indicates deeper water and muck soils. Venturing into these areas can be a rude surprise, resulting in stuck feet and a wet wallet (sometimes even if it is in a shirt pocket). There may also be another, possibly dangerous surprise. These plants frequently fringe alligator holes. This thought can be disconcerting while you are struggling to catch your balance and your feet are stuck in the muck!

Periphyton[22, 224]

Although rooted wetland plants dominate the appearance of the Everglades, algae may be the more important producers at the base of the food chain. Algae

---

* The term *refugia* (plural of the Latin *refugium*) is used here to avoid confusion with various uses of the word "refuge," such as national wildlife refuge. In Everglades jargon, refugia are habitats where aquatic organisms survive temporary drydowns of Everglades marshes, especially solution holes and alligator holes, but also sloughs in years when they remain continuously flooded.[133]

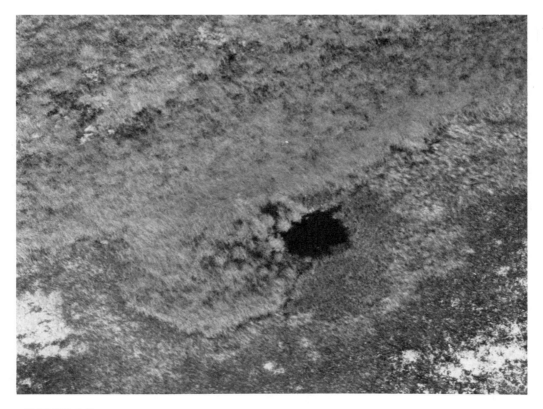

FIGURE 3.9
Aerial view of an alligator hole. This example, about 15 feet across, was in the Shark River Slough, Everglades National Park. The community above and to the left of the hole is sawgrass marsh, while wet prairie on peat lies below and to the right. The light areas are periphyton. (Photo by T. Lodge.)

in the Everglades, as in aquatic habitats everywhere, grow on any submerged surface wherever adequate sunlight is available. Plant stems and the soil surface of the marsh are covered with a complex association of numerous types of algae in a community called *periphyton* (Figure 3.10). Where it grows on the soil surface, it is commonly called *algal mat*. Snails and herbivorous fishes graze on Everglades periphyton, and it is considered to be a very important food source.

Two distinct types of periphyton occur in the Everglades. In more acid or soft water, with little dissolved calcium and which normally occurs over peat soils in the slough and deeper wet prairie communities, the periphyton is a thin association of mostly green algae. This type has been shown to be a nutritious food for grazing fishes and invertebrates, such as flagfish, sailfin mollies, prawns, and applesnails.

In areas of harder water, which contain abundant calcium from contact with limestone bedrock, the periphyton looks and feels very different. It contains relatively more blue-green species of algae, is usually a light yellowish-brown,

FIGURE 3.10
Periphyton surrounds plant stems in marl prairie, near Pay-Hay-Okee Overlook, Everglades National Park. (Photo by T. Lodge.)

and forms a thick mass with a spongy texture. Around thin plant stems, it may be a half inch in diameter, and on the soil perhaps an inch thick. The texture is due to minute calcium carbonate crystals that are formed as the algae remove carbon dioxide from the water during photosynthesis. This type of periphyton is also consumed by grazers, but is not as nutritious as the other type, possibly because of the calcium carbonate particles (which make it "crunchy"!). It typically occurs in areas of short hydroperiod, notably in the wet prairie community on marl soil.[84, 135]

Periphyton has been shown to be sensitive to changes in water chemistry and nutrients, as well as other ions, as evidenced by shifts in the types and diversity of algae. Typical periphyton in the Everglades is adapted to water containing low levels of nutrients, and the addition of nutrients changes the assemblage. Certain types, notably species of blue-green algae, can "fix" nitrogen, and they become more common when the water is enriched in phosphorus.

Marsh Soils [34, 39, 84, 84a]

Two types of soil occur in the Everglades: marl and peat. Both were mentioned previously in the context of the plant communities. Marl is a product of periphyton. During the dry season, the organic material (dead algae) in the periphyton mass oxidizes, leaving the calcium carbonate particles as this light-colored soil. Marl is descriptively called *calcitic mud* and is the main soil of the short-hydroperiod wet prairie habitats near the edges of the southern Everglades, where the bedrock lies close to the surface. It is the primary soil type along much of the main road to Flamingo in Everglades National Park.

The "opposite" type of soil, peat (and related muck), usually occurs in areas where the bedrock lies deeper. Peat is composed of the organic remains of dead plants. Examination (requiring magnification) can reveal the kinds of plants involved if the soil has remained undisturbed.* However, if the material has become too finely divided (often due to animal activity by everything from earthworms to alligators), the soil is commonly called muck. This non-technical term also applies to black peat soil combined with non-organic material.[23] The Everglades soil near Lake Okeechobee contains substantial sediments originally from the lake and is thus called muck.

Coal is actually fossilized peat and muck, and coal even hundreds of millions of years old often contains fossil imprints of the plants that flourished in those ancient wetlands. Most of the peat in the Everglades is either decomposed sawgrass or slough vegetation, and it occurs where the historic hydroperiod was very long, normally keeping the soil saturated throughout the year. This condition occurred historically in sloughs, long-hydroperiod wet prairies, and sawgrass marshes of the northern Everglades.

Water is required for the formation of peat soil because it restricts atmospheric oxygen from the soil. Without oxygen, microorganisms cannot decompose dead marsh plants as fast as they accumulate. Through the years, ever thicker deposits of peat are deposited until an elevation is reached where the surface dries enough that oxygen-dependent decay (the opposing aerobic process), or fire, prevents accumulation.

Near the turn of the century in the northern Everglades, the peat and muck

---

* Two types of peat commonly found in the Everglades marshes are sawgrass peat and Loxahatchee peat. Sawgrass peat is the more widespread, especially in the northern and central Everglades. Loxahatchee peat occurs in deeper marsh areas, typically the northeastern Everglades and the Shark River Slough. It contains less sawgrass remains and more evidence of slough vegetation such as the roots, rhizomes, leaves, and seeds of water lilies.

soils were commonly 14 or more feet deep over the rock base, when drainage activities began. In the southern Everglades, peat soils were much thinner, but in the deepest parts of the Shark River Slough there were (and still are) several feet of peat. Look back for a moment to the northern Everglades, where most of the area has been developed for agriculture. Those soils were coveted because they were enriched by sediments from Lake Okeechobee, but agricultural drainage and soil tilling practices stimulated their loss. Through the decades, these soils, which supported some of the most productive agriculture in the world, literally disappeared. This process, called *subsidence*, has resulted from compaction (by drying), from fire, and from aerobic decay. Like coal, dried peat will burn, and fires have been responsible for substantial soil losses. However, the most persistent factor is aerobic decay by microorganisms that degrade the peat into carbon dioxide and water. In large areas, more than eight feet of soil depth has been lost. Visual evidence is provided by houses originally built on pilings, but with their frames located on then existing ground surface. Years later these houses stood high in the air, requiring long stairways to reach the doors.[23, 221]

Marl and peat soils are opposites, requiring aerobic and anaerobic soil conditions, respectively. Peat cannot accumulate in shorter hydroperiod marshes where marl persists, and acid conditions within a peat soil dissolve marl, preventing its accumulation. However, there are large areas in the northern Everglades where a thick layer of marl *underlies* peat. This occurrence stands as evidence that the hydroperiod increased substantially in those areas as the Everglades ecosystem evolved.

Water Quality[46a, 211]

The appearance of the water in the Everglades marshes surprises many first-time visitors. They frequently expect the water to be unpleasant in some way (perhaps turbid and foul smelling), but it is clear. The water ranges from very soft and slightly acidic where it is derived from rainwater that has only contacted vegetation and peat soils to hard and alkaline (containing abundant dissolved calcium and other ions) where it has contacted the limestone bedrock (see the section on periphyton earlier in this chapter). Changes in Everglades water quality have become highly controversial. Relevant issues including urban and agricultural activities, water routing by canals, and even man-altered rainwater chemistry are addressed in Chapter 18. The "original" conditions are covered here.

Historically, the water of the central and southern Everglades has been very low in the nutrients that promote plant growth, including nitrogen and phosphorus, which are principal components in fertilizers. The Everglades has been described as a "nutrient-limited system" and very low phosphorus, at or below 10 parts per billion total phosphorus, has been indicated as the main limitation. In contrast, the historic levels of nutrients in Lake Okeechobee were much higher than in the Everglades, apparently because of natural sources of phosphorus within its watershed. It is probable that the great original expanse of dense, tall sawgrass of the northern Everglades, south of Lake Okeechobee, was stimulated by nutrients from the lake's overflow waters. In turn, that sawgrass area extracted the nutrients, enriching the northern Everglades soils and leaving the lower Everglades with the very low-nutrient waters. The main source of nutrients for the central and southern Everglades was probably rainfall.

The characteristic Everglades marsh plants thrive in this low-nutrient environment, but this condition accounts for the rather sparse, open character of much of the marsh. In turn, the open character is conducive to the kinds of aquatic animals (fish and invertebrates) that are present. The sparsely vegetated surface promotes adequate gas exchange with the air so that the water is normally well oxygenated. In marshes with nutrient-rich waters, other kinds of vegetation predominate, sometimes in such thick profusion that little open water is available. Such situations promote poorly oxygenated conditions and do not provide good habitat for most amphibians and fishes, although certain aquatic insects and other invertebrates may be abundant.

Cattails are among the marsh species that compete successfully where there are higher levels of nutrients or where disturbance has removed other marsh plants. In the Everglades, cattails occur naturally only where soil fires removed other marsh plants or in and around depressions such as alligator holes.* Soils in alligator holes contain elevated levels of phosphorus deposited by the aquatic life that perishes during severe dry seasons.[131] Together with excrement from wading birds and alligators that feed in these areas during low water, the bodies of these aquatic organisms act as fertilizer, enriching the soil and creating a localized environment for different marsh plants.

The final word on Everglades water is that it is directly connected to the water supply of Florida's urban southeast coast. The same highly porous limestone bedrock is continuous from the central and southern Everglades to these areas, a renowned conduit known as the Biscayne Aquifer.[114]

Weather and Fire[31, 39, 92, 101, 245]

Any study of the Everglades requires an understanding of the role of fire, without which the Everglades would be very different. It would probably be much more of a swamp, dominated by wetland trees. Fire is an important controlling agent for most habitats of southern Florida, as it is in most seasonally

---

* Of three species of cattail that occur in the Everglades region, two (*Typha domingensis* and *T. latifolia*) have invaded marshes presumably due to nutrient enrichment.[191] They do not, however, normally occur even in the deeper marsh communities, such as sloughs, where the deeper water would seemingly be conducive to their invasion; they tolerate deeper water than sawgrass.[3] They are mentioned only as a minor element in descriptive literature of natural Everglades plant communities. Certain species such as spike rush are known to inhibit cattail establishment.[126] Cattails are well known for their rapid invasion of a wide variety of disturbed habitats (even flat roofs of buildings) and are often cited as a problem in wetland mitigation.[42, 62, 165]

dry areas of the earth. Fire prevents the invasion of the grassland by shrubs and trees, a process called *succession,* and has the positive effect of releasing and redistributing nutrients that become "locked-up" in plant tissues during growth. Death and decay would eventually release the nutrients, but fire is faster. Thus, in weeks following "normal" fires, new growth rejuvenates the marshes. However, defining "normal fire" requires an understanding of the region's seasonal cycles of weather.

The average annual rainfall in the Everglades varies between 50 and 60 inches, depending upon location, with significantly more falling along the eastern edge. As a long-term average, three quarters of the annual total comes during the six-month wet season from May through October.[144] Water levels typically rise very rapidly at the beginning of the wet season, but do not reach their maximum until late in the season. Trade winds influence the weather during the warm summer months, bringing a continuous supply of warm, moist air westward over the Atlantic, and localized afternoon thunderstorms develop daily. Of course, the rainfall is sometimes increased catastrophically by tropical weather events such as depressions, storms, and occasional hurricanes.

It is during the summer rainy season that nature's fire starter—lightning—plays an important role, and the summertime frequency of lightning in peninsular Florida is very high.[164] The fact that lightning starts wildfires is well documented;[197] in an eye-opening demonstration, a lightning strike made a cabbage palm seemingly explode into flame.[131a] Summertime fires usually do not damage soil or roots, because summer rainfall increases soil moisture and water levels. Thus, the dead portions of the Everglades vegetation burn but little damage is done.

Now consider the six-month, winter-spring dry season when only one quarter of the annual rainfall occurs. Late October or early November nearly always bring the first abrupt change in weather after a long, hot, wet summer. The arrival of the first cold front is a pleasant relief, at least for the human residents. In mid-winter, some of the cold fronts are unpleasantly cold, but the mild first arrivals do little more than signal the change in dominant weather from the tropical trade wind system to the temperate, prevailing westerly system. The latter is characterized by "frontal" weather, pulsing a new cold front through the region almost weekly through most of the winter. As these air masses move down the peninsula, they warm, reducing the relative humidity. In short, southern Florida "dries out," and water levels in the Everglades typically recede gradually, but with minor increases due to sporadic dry season rain events.*

Most cold fronts in southern Florida pass with little or no rain, just some windy weather followed by much cooler, cloudless sky. Cold fronts are rarely

---

* Substantial deviations from the normal dry season regime occur as a result of the El Niño phenomenon in the Pacific. Unusually warm sea-surface temperatures in the Pacific Ocean bring abnormally wet weather to the southern United States, including Florida, during the winter (early and middle dry season).[68, 82] The winter of 1982–83 and the three consecutive winters of 1991–92 through 1993–94 (the longest El Niño on record) were exceptionally wet because of El Niños,[161] with only a short, late dry season occurring during those years.

violent enough to have electrical activity, and the few violent fronts are almost always wet, at least initially. A fire started by lightning at such times is likely to be extinguished by widespread rain, unlike the localized summertime thunderstorms, which may pour rain on one spot and provide nothing but some frightening bolts of lightning at the next.

Dry season fires can change Everglades landscapes extensively, especially during severe droughts, when peat soils are dry. At those times, both roots and soils go up in smoke, along with the rest of the plants. However, the earliest thunderstorms of the rainy season, in April or May, can start enormous fires in the still dry Everglades. Such catastrophic fires have always been part of the natural regime, and their charred evidence can be found in buried layers of peat soils.[33] These fires reduced accumulated peat levels, preventing succession and perpetuating long-hydroperiod marsh habitat. They occurred infrequently, however, perhaps once in a decade, and in the long intervening periods, Everglades wildlife prospered. Since the early 1900s, however, man's unmanaged dry season fires—due to arson or accident—have become a problem of great magnitude.* On the other hand, fire is now used by wildlife officials to manage many natural areas. Prescribed burns in the Everglades are now an important tool in an evolving science that strives to mimic the original, natural effects of fire.

---

* It is assumed that Native Americans in the Everglades region used fire, primarily as an aid in hunting, and the region was altered as a result. However, whatever impact they had still resulted in a ecosystem that provided varied habitats, including fire-sensitive plant communities and abundant wildlife.[197]

4

Wetland Tree Islands*

T he term *tree island* is derived from the island-like appearance of a small forest situated in an open area such as a marsh. The term is not unique to the Everglades, but it is certainly appropriate because certain areas are dotted with thousands of them. Tree islands add considerably to the functional value of the surrounding marsh by providing habitat diversity. They provide nesting sites for alligators, protective cover for wildlife such as deer and otters, and roosting and nesting locations for birds, notably wading birds.[74, 131b]

There are two very different types of tree islands: those composed of species that regularly grow in wetlands and those composed of "upland" species, which only grow on land above normal flooding elevations. Upland tree islands are more properly called *hammocks* and are covered in Chapter 5. In contrast to tropical domination in the hammocks (at least those of the southern Everglades), most of the wetland tree island species are of temperate, North American origin and are common in swamps throughout the southeast. Common wetland tree island species found in the Everglades are listed in Table 4.1.

Tree islands range from fractions of an acre to hundreds of acres. The larger examples are found in Shark River Slough, where they form an interesting elongated shape, oriented with the flow of water and often described as "teardrop." The upstream end sometimes has an area of elevated rock that supports a hammock. Around the area of higher elevation, and forming the long, downstream tail, are the island's wetland trees. Given their orientation and their appearance from the air, "comet-shaped" would be an appropriate description for some, but the comet "tail" is dominated by dense sawgrass.

Types of wetland tree islands are named for the kinds of trees that dominate them. The most common are dominated by red bay or sweet bay and are called *bay heads* (sometimes one word). Those dominated by coastal-plain willow are called *willow heads,* and those dominated by cypress, *cypress heads.* Other species

---

* The primary citations for this chapter are References 39, 46, 92, and 175.

TABLE 4.1

Wetland Tree Islands: Common Trees, Shrubs, and Understory Plants of the Everglades[a]

Common name	Scientific name
trees	
bald cypress	*Taxodium distichum*
pond cypress	*Taxodium ascendens*
red bay	*Persea borbonia*[b]
sweet bay	*Magnolia virginiana*
red maple	*Acer rubrum*[c]
small trees/shrubs	
pond apple	*Annona glabra*
dahoon holly	*Ilex cassine*
coastal-plain willow	*Salix caroliniana*
cocoplum	*Chrysobalanus icaco*
wax myrtle	*Myrica cerifera*
buttonbush	*Cephalanthus occidentalis*
understory and ground cover plants	
leather fern	*Acrostichum danaeifolium*
swamp fern	*Blechnum serrulatum*
Boston fern	*Nephrolepis exaltata*
royal fern	*Osmunda regalis*

[a] Selection of species is based on information provided by Craighead,[39] Loveless,[140] and Olmsted and Loope.[175]

[b] Following Tomlinson,[236] the name *P. borbonia* is recognized here as including *P. palustris*. In other taxonomic literature (i.e., Long and Lakela[139]), *P. palustris* (swamp bay) is recognized as a separate species, but most of the descriptive literature of the Everglades lists only *P. borbonia*.

[c] Red maple is a good example of the reduced temperate flora in extreme southern Florida. Common in tree islands of the northern Everglades, its appearance declines southward and is rare in Everglades National Park.

of wetland trees commonly found in wetland tree islands include cocoplum, dahoon holly, wax myrtle, pond (or custard) apple, and buttonwood, but these species are not used as names for types of tree island, even if they occasionally dominate.

Bay Heads[47, 84, 84a, 140, 245]

Bay heads (Figure 4.1), by far the most common type of tree island, often occur over discrete depressions or cavities in the limestone bedrock that are filled with peat soil to an elevation of one to three feet above the surrounding marsh. The hydroperiod, shorter than in the surrounding marsh, typically ranges from two

FIGURE 4.1
Tree islands (bay heads) contrast abruptly with slough and wet prairie vegetation in the Shark River Slough, Water Conservation Area 3A. A narrow moat is visible at the left end of the nearer island. Similar bay heads are abundant in the Arthur R. Marshall Loxahatchee National Wildlife Refuge (Water Conservation Area 1). (Photo by T. Lodge.)

to six months. Bay heads are not uniformly distributed through the Everglades, but are most common in the northeastern or Hillsborough Lakes area (now the Arthur R. Marshall Loxahatchee National Wildlife Refuge or Water Conservation Area 1), the west-central portion, and throughout the Shark River Slough in the southern Everglades. These areas are typified more by slough communities rather than shallower marshes. The long hydroperiod of the surrounding marsh often protected tree islands from fire, possibly accounting for their distribution.

The origin of bay heads has puzzled many investigators. Theories have included natural, in-place succession from a previous marsh community; dislocated, floating blocks of marsh peat with subsequent invasion by trees; and remnant areas surviving from a hypothetical swamp forest over the entire Everglades. Investigations of the soil beneath bay heads sometimes reveals tree island (woody) peat—called *Gandy peat*—on top of peats of marsh origin, but often Gandy peat extends down to the underlying rock. Thus there may be more than one explanation for the origin of tree islands, but persistence from a once widespread swamp is not supported. In fact, recent studies indicate that bay heads began to appear in the Everglades only about 1000 years ago and that the landscape was devoid of these vital features in its earlier history.

Around the perimeter of many bay heads is a zone, ranging from 10 to 20 feet wide, that is both deeper and less densely vegetated than the surrounding marsh. Often called *moats,* these areas are heavily used by alligators, fish, and other aquatic life. One or a combination of two factors may explain the existence of tree island moats. In areas of marl soil, acids from decay of litter in the tree islands are thought to dissolve marl around the edge of the tree island. However, regular use by alligators is another factor. It seems logical that alligators would benefit from regular patrols of the edges of tree islands, where the chance of ambushing prey would be far higher than in the open marsh. To extrapolate further, the moats might often have the effect of preventing marsh fires from entering tree islands. If the "alligator theory" is even partially correct, then alligators also help to maintain the integrity of bay heads.[38, 39]

Some bay heads may have been started by alligators. Wetland trees can first take root where an alligator has piled soil while excavating a pond in a marsh. After years of root and litter accumulation, in addition to further efforts of alligators in piling soil, an alligator hole would look like a tree island from the outside, but from the air would look like a ring of trees around a central pond (Figure 4.2).

Alligators commonly associate with existing bay heads. The elevated peat soil is ideal for constructing both dens and nests, with the pond at the edge. Following the sequence described earlier, a tree island could eventually enclose an alligator hole started in this way. Alligators are also known to take advantage of bay heads that have been damaged or destroyed by fire during an extreme drought, to the extent that a depression is formed where the peat soil burned. Use of such a naturally provided pond would similarly allow a ring of trees to develop. By whatever means, the reality is that alligator holes are often located within wetland tree islands, and alligator activity is probably important in the perpetuation and integrity of tree islands.

Willows and Willow Heads

Coastal-plain willow, a low tree that colonizes rapidly with wind-blown seeds, may form discrete willow heads but is associated with a variety of disturbed habitats. It prefers lower, wetter elevations than most species of the bay head community and is a good marker for an active alligator hole. Willow heads often mark the location of a bay head that has been completely destroyed by fire, especially where fire has burned the peat soil, reducing the elevation. Willows have also overtaken extensive areas of the Everglades where the combination of reduced water level and fire has eliminated sawgrass. However, in the long term, without fire, willows are typically displaced by other tree species.

FIGURE 4.2
An alligator hole in the edge of a bay head (without a moat), Shark River Slough, Everglades National Park. (Photo by T. Lodge.)

Cypress and Cypress Heads[64]

Cypress trees are found in depressions where the hydroperiod is usually longer than in the surrounding marsh. Two species of cypress—bald cypress and pond cypress*—occur in the Everglades, but in very limited areas. Historically, a narrow cypress swamp bordered the eastern edge of the Everglades in Palm Beach and northern Broward counties, south to the New River at Fort Lauderdale, and small vestiges of this former community are scattered through developed areas of those counties. Examples are found along Sample Road

---

* Some authorities consider bald and pond cypress to be two forms of the same species.[64, 236]

FIGURE 4.3
A small cypress dome in winter, leafless condition, along the main park road to Flamingo, just north of the entrance road to Mahogany Hammock. (Photo by R. Hamer.)

between State Road 7 and Rock Island Road in northern Broward County and at the visitor's center of the Arthur R. Marshall Loxahatchee National Wildlife Refuge in Palm Beach County, in the former headwaters of the Loxahatchee Slough.

The occurrence of cypress in the Everglades is now restricted to limited numbers in Loxahatchee National Wildlife Refuge and to areas in Everglades National Park near the pinelands, east of the Shark River Slough and into Taylor Slough. In the latter area they form a very sparse, open forest of stunted trees, often called *hatrack cypress*, which may be hundreds of years old, yet seldom taller than 25 feet. More like marshes than swamps, these areas, called *dwarf cypress forests* or *cypress prairies*, are eerie in appearance but subtly beautiful (see Figure 3.5). In deeper pockets, the trees grow better and form dense *cypress heads*, composed almost entirely of cypress (Figure 4.3). Unlike their pine relatives, cypress are deciduous, shedding their small leaves each fall. Their light, grey-brown bark gives cypress heads and cypress prairies a stark, bizarre appearance in winter, in contrast to their beautiful, light-green summer foliage.

Within cypress heads, growth conditions are better toward the center. If the area is large enough (several acres or more), cypress heads form a very distinctive dome shape and are called *cypress domes*, with progressively taller trees toward the center. The peat soil that forms in these low areas of long hydroperiod is more favorable to cypress. Nevertheless, the trees do not reach heights of more than about 50 feet in the Everglades, while the same species grow much larger in the neighboring Big Cypress Swamp. Most of the Big Cypress Swamp area was logged in the 1950s, but some very large cypress remain in Corkscrew Swamp, a sanctuary maintained by the National Audubon Society.

The roots of cypress trees produce growths that protrude above the soil and look like sawed-off trees that somehow healed over. These structures, called *knees*, may range from a few inches to more than six feet tall. Cypress knees are thought to be "breathing organs" for the roots, where gases are exchanged with the atmosphere. Because the soils where cypress normally grow are usually saturated with water, they are deficient in oxygen. Roots require oxygen, just as other plant tissues do. Only the leaves produce oxygen, and then only during daylight. Thus, the knees are an adaptation to water-logged or hydric soils. Such adaptations are also important among mangroves (see Chapter 7).[236]

Another characteristic of cypress is fire resistance. Like pine trees, cypress can survive fire with little apparent damage. However, fire can be decimating during an extreme drought if the soil has dried. Dry peat will burn, causing root damage that kills the trees. Cypress trees can grow in uplands, but fire restricts their normal distribution to perennially saturated soils. Cypress heads that have been damaged or destroyed by fire are often invaded by willow.

Pond Apple/Custard Apple

The most peculiar tree associated with tree islands is the pond apple (Figure 4.4), a species that prefers a longer hydroperiod than most. The edge of an alligator hole, the moat on the outer edge of a bay head, or a cypress head interior are ideal for the pond apple. Though seldom taller than 35 feet, it develops a massive trunk and large, buttressed roots. Together with the robust load of epiphytes that typically thrive on its rough, dark-brown bark, it presents a bizarre spectacle. Due to its preferred location next to open water and its convenient configuration of branches, the pond apple is regularly used by nesting water birds, such as the anhinga. The greenish-yellow to yellow fruit resembles a large apple and is eaten by animals, which then serve to distribute the seeds. Most people describe the taste as sour (thus the alternate name *custard apple*) with an unpleasant musky odor. Politely stated, it is undesirable.

Historically, a forest of pond apple existed along the south shore of Lake Okeechobee, at the "head" of the Everglades (see "The Historic Everglades" in Chapter 2). That pond apple swamp—so different from the remainder of the Everglades—may have flourished because of nutrients and fine sediments from Lake Okeechobee, and its legacy was embodied in its rich soils (termed Okeechobee and Okeelanta Peaty Mucks). Like the sawgrass peat soils of the northern Everglades, these soils were very desirable for agriculture, and the swamp was converted to cropland in the early decades of this century.[127a, 221]

FIGURE 4.4
Leafless through part of the dry season, a pond apple supports innumerable air plants, primarily stiff-leaved wild pines at the Anhinga Trail, Everglades National Park. (Photo by R. Hamer.)

Role and Integrity of Tree Islands[6, 38, 74, 140, 245]

Tree islands greatly enhance the ecological value of the Everglades, especially where they support wading bird rookeries and harbor alligator holes. Their survival depends on a delicate balance between fire and water levels, and the alligator is undoubtedly involved. Hydrologic alteration of the Everglades, and probably the reduced numbers of alligators, have caused many complex changes in tree islands: enlargement, shifts in the kinds of trees that dominate them, or even complete destruction.

5

Tropical Hardwood Hammocks*

A hammock is a localized, mature hardwood forest. By *hardwood*, the distinction is made that broad-leaved trees are prevalent, as opposed to pines, which normally have softer wood. The term *hammock* is of uncertain origin, but it may have been derived from a Seminole Indian word meaning house or home, and many hammocks were used by Seminoles as home bases. In southern Florida, hammocks occur in marshes, pinelands, mangrove swamps, and the interiors of some wetland tree islands. In order for hammocks to exist, the ground surface must be high enough so that seasonal flooding does not occur.

Hammocks sometimes occur as tree islands in Everglades marshes, but they are in marked distinction to the wetland tree islands, which are dominated by a temperate swamp flora. In the northern Everglades region, hammocks may be dominated by trees of temperate climate origin such as the live oak or hackberry (*Celtis laevigata*), and the occurrence of tropical trees is much more restricted. From the latitude of Miami southward, however, nearly all of the trees are of tropical origin, with live oak being the only significant temperate representative. For this reason, the term *tropical hardwood hammock* is commonly used for any hammock in Everglades National Park, the Florida Keys, or the Miami area.

Early botanists visiting southern Florida were impressed by the tropical hardwood hammocks.[78, 219] The types of trees in these hammocks have unusual names for anyone not familiar with the West Indian region of the American tropics. Some of the prevalent trees and shrubs in the hammocks of Everglades National Park are listed in Table 5.1. Of these, only the live oak originated in temperate North American. All others are of tropical origin, and the seeds of nearly all of them are dispersed by birds, which explains how they probably got to Florida

---

* The primary citations for this chapter are References 41 and 175.

TABLE 5.1
Common or Well-Known Trees and Shrubs of the
Tropical Hardwood Hammocks of Everglades National Park[a]

Common name	Scientific name
Florida royal palm	*Roystonea elata*
cabbage palm	*Sabal palmetto*
live oak	*Quercus virginiana*
strangler fig	*Ficus aurea*
pigeon plum	*Coccoloba diversifolia*
lancewood	*Nectandra coriacea*
wild tamarind	*Lysiloma latisiliqua*
paradise tree	*Simarouba glauca*
gumbo limbo	*Bursera simaruba*
West Indian mahogany	*Swietenia mahagoni*
poisonwood	*Metopium toxiferum*
inkwood	*Exothea paniculata*
white stopper	*Eugenia axillaris*
marlberry	*Ardisia escallonioides*
wild mastic	*Mastichodendron foetidissimum*
wild coffee	*Psychotria undata*

[a] Selection is based primarily on data supplied by Olmsted and Loope[175] (Table 2, page 180 therein), but with the addition of some uncommon species for general interest.

from the West Indies where they occur naturally. The only listed species that is not common is the Florida royal palm, which is included because of its magnificent stature. It is closely related to the Cuban royal palm; both have massive, light grey trunks that look like concrete. Wherever royal palms grow, they usually reach conspicuously above the other hammock trees.

The Hammock Environment

The feeling inside a tropical hardwood hammock is special. For those familiar only with temperate forests, the hammocks are full of strange sights and sensations. A first impression is the smell, which is unlike any northern forest. It reminds most people of a faint skunk odor, yet is not unpleasant. This aroma is emitted by one of the most ubiquitous of the hammock species, a small tree called the white stopper, named "stopper" for its use in treating diarrhea. The protection afforded by the cover of trees also creates acoustic effects, with an eerie silence prevailing.

Because most of the foliage is overhead, the inside of a mature hammock is easy to explore, with the dense shade preventing much growth at ground level. The forest floor, on the other hand, may present obstacles. There are numerous rough limestone rock outcroppings and frequent solution holes or "sinks," some of which are deep enough to contain a pool of water. In general, hammocks developed where the surface rocks were slightly harder than in surrounding areas. This condition resulted in slower weathering of the rock, leaving the elevation higher. However, chemical weathering of the limestone underneath these areas continued until, here and there, the surface collapsed, producing a sink. Exposed ground water within the sinks contributes to higher humidity, promoting luxuriant growths of mosses and ferns on the walls. Most of these species are also of tropical origin. More ferns, such as the resurrection fern, are found on the trunks and limbs of the hammock trees, in addition to air plants and orchids.

The dense shade provided by the trees of a mature hammock and the exposed ground water help keep the inside temperature several degrees cooler during hot summer weather. Conversely, during severe winter cold fronts, tropical hammock plants are usually spared by the higher humidity and protection from wind, while the same kinds elsewhere perish.

Tree Height

Although the trees protect plants inside the hammock from cold weather, the tallest trees are frequently damaged at their tops and effectively pruned by the cold. The cold, together with hurricanes and lightning, explains why hammock trees do not grow very tall, usually not reaching more than 50 feet. Any tall tropical hammock tree in the path of a hurricane is sure to be uprooted, because the roots of these species are surprisingly shallow (Figure 5.1). The summertime incidence of lightning strikes in peninsular Florida is very high,[164] and the tallest trees are prime targets. Without these limitations, most of the tropical hardwoods would be much taller.

The Strangler Fig

One of the common hammock trees is so strange in its growth form that it deserves special recognition: the strangler fig (Figure 5.2). There are two native figs, both from the West Indian tropics, but the strangler is more common. This species produces a berry-like fruit commonly eaten by birds. The seeds within the fruit are not digested and remain viable for dispersal in the birds' droppings. Although the seeds can start growing wherever there is sufficient moisture, one of the most common places is in the "head" of a cabbage palm. Thus, a fig seed sprouting there begins life as an epiphyte. Like orchids, ferns, and air plants, figs

FIGURE 5.1
Chaos at Royal Palm Hammock five months after Hurricane Andrew. A gumbo limbo rests against another, and vines have advanced over much of the undergrowth and up many of the trees. (Photo by T. Lodge.)

FIGURE 5.2
An offshoot of a strangler fig (defoliated, winter condition) encircles a dead, deteriorating cabbage palm. The main trunk of the fig stands at left and a cactus grows from the top of the palm. Snake Bight Trail near Flamingo, Everglades National Park. (Photo by R. Hamer.)

are not parasites because they do not derive nutrients from the host tree, but the end result is the same. As the fig grows, it proliferates roots downward, toward the forest floor. On reaching soil, they expand greatly in size and form a network that completely entwines the host, which is eventually killed by the fig tree's dense shade. The strangler fig literally takes the host's place in the forest. Strangler figs are not choosy; they engulf stone walls, abandoned cars, and even household chimneys if not removed.

Hammocks and Fire

A common characteristic of all tropical hardwood hammocks is intolerance to fire. Fire actually helps maintain many other kinds of plant communities in the Everglades, especially marshes and pinelands, but it can destroy hammocks. Fortunately, the very characteristics that make hammocks such interesting places also help prevent fire from entering. These characteristics are the open understory, with little accumulated fuel, and the high degree of shade, which keeps the inside climate relatively cool and moist. In very dry years, however, hammocks are highly vulnerable to fire, which may then destroy the humus soil. Soil fires burn slowly but persistently, killing the trees by root damage. Hammocks may survive a surface fire and look almost untouched after a few years, but soil fires have completely eliminated many hammocks.

Unpleasant Aspects of Hammocks

So far, a wondrous picture of the tropical hardwood hammock has been painted here, but there are also drawbacks. First is the problem of just getting inside. While it is easy to walk around inside a mature hammock, it is not easy to enter it from the outside. The edge of a hammock does not have the tall trees that produce dense shade under which little grows at ground level. On the contrary, the edge is nearly impenetrable. The outermost edge is sometimes dominated by the saw palmetto (*Serenoa repens*), a low, sprawling species of palm which is common in the pinelands. This species seldom grows over about six feet tall, and its trunks lie prostrate along the ground. The leaf stem (properly called the petiole) is armed with upward-pointing teeth that make those of sawgrass look and feel like child's play. The saw palmetto's stiff, toothed petioles can cut legs and ankles and the trunks can make walking difficult, while the large broad leaves may hide impending dangers. This exciting adventure is intensified by the possibility of encountering the sometimes huge (up to about seven feet long) eastern diamondback rattlesnake that may lie concealed among saw palmettos. The species is not common, but the thought is still frightening.

The area of saw palmettos is generally the easy part, and they are not even present at all hammock edges. The next rim of fringing vegetation, which is always present, is a tangle of shrubs and vines that defy intrusion. In addition to this difficulty, one of the most common species in this fringe is poisonwood, a close relative of poison ivy which similarly produces an itchy, unpleasant skin rash. The leaves of poisonwood are deceptively similar to those of gumbo limbo, a harmless hammock species, and many people are fooled by the similarity. Once through the hammock's edge, you have paid the price and are free to walk through the hammock forest. Fortunately, many hammocks have been "conquered" for public access with pathways and boardwalks. Royal Palm Hammock and Mahogany Hammock in Everglades National Park are prime examples.

The second unfortunate feature for visitors to a hammock takes only one word to describe: mosquitos! Mosquitos can be unbearable under the shade of hammock trees, particularly near salt water, during the warmer, wetter months of the year, generally May through October. If a hard-blowing cold front has not occurred for a month or so, mosquitos may be a problem even in winter months. The answer is to use plenty of insect repellent.

Hammocks and Wildlife

Hammocks that occur inside wetland tree islands, or alone in an area of marsh, may support a wading bird rookery. However, most hammocks accessible to people lie within other upland communities and would not normally be used by colonial nesting birds, which seek the protection afforded by surrounding water. Thus, there is an element of disappointment for many visitors who expect to see abundant animal life. Hammocks are not a showcase for wildlife. Some of the interesting Florida tree snails, a few lizards (most commonly the southeastern five-lined skink, a sleek lizard with an electric-blue tail), occasional owls (barred owls are the most common), and some raccoons can generally be seen, but more often, only patient, skilled observers are rewarded. Hammock wildlife is not rare; it is only difficult to see. The spectacular vegetation is the main attraction.

6

Pinelands*

Many visitors to Everglades National Park are surprised to find that one of the most extensive habitats appears to be the open pineland, commonly called *pine flatwoods* or simply *flatwoods.* The dominant species, and the only true pine present, is slash pine (*Pinus elliottii*), a handsome, long-needled species. The name *slash* is derived from the once widespread practice of extracting sap from this species by cutting diagonal slash marks in the trunk. Sap draining from the cuts was collected in cups and used to make turpentine and other products.

Slash pine is abundant throughout Florida and neighboring areas of the southeast, where it is the most important species in the paper (pulp) industry. In northern Florida enormous acreages of slash pine are planted. In southern Florida, a special variety of the species—*Pinus elliottii* var. *densa*—is locally known as Dade County pine.[236] Older *densa* variety trees have stored unusually large amounts of resins, which makes the wood very hard (dense), unlike most pines. The early lumbering industry of the Miami area (Dade County) used this once abundant resource, and houses made with this wood now have a time-tested value: resistance to termites. The only drawback to the wood is that it is difficult to drive a nail into it.[39]

Slash pine prospers in a range of elevations from "high and dry" to levels that are only slightly (inches) above marsh habitats, and shallow flooding may occur for two or even three months during the height of the summer rainy season. Pinelands originally dominated the Atlantic Coastal Ridge, upon which most of the metropolitan areas of Miami and Fort Lauderdale have been built. The southern part of this ridge, from Miami southward, is made up of very rough limestone, full of holes and cavities. It provides a very different ground surface for the local "rocky" pinelands, compared to the sandy soils common to the flatwoods that border the northern Everglades and in the remainder of the state. The "tail" of this ridge curves westward into Everglades National Park and extends in "bits and pieces" to Mahogany Hammock, one of the best known of

---

* The primary citations for this chapter are References 1 and 217.

FIGURE 6.1
Rocky pinelands in Everglades National Park near the entrance from Florida City. Low shrubs including saw palmettos and cabbage palms with an open understory characterize the appearance of pine flatwoods throughout the Florida peninsula. (Photo by T. Lodge.)

the park's tropical hardwood hammocks. The main road through Everglades National Park follows this rocky pineland habitat for many miles, giving the impression that much of the park is pineland (Figure 6.1). Actually, the pinelands occupy only a small percentage of the total mainland area of the park.

Pinelands and Fire

The pineland is termed a "fire climax" by ecologists, which means that pineland is the final vegetation that develops (on acceptable elevations) if fire is a regulat-

FIGURE 6.2
Fire moves through rocky pinelands near the entrance road to Mahogany Hammock, Everglades National Park. Its impact is important in maintaining the integrity of this community. (Photo by R. Hamer.)

ing factor (Figure 6.2). Slash pine and its associated vegetation are highly resistant to fire and are actually maintained or perpetuated by fire. Aside from various grasses and a host of other very interesting small pineland plants (many of which are of tropical origin and only found in the rocky pinelands of southern Florida), two very obvious species in the pineland community are palmettos. The saw palmetto (*Serenoa repens*), a small, low-growing palm species which was described in Chapter 5, is the most abundant. The taller cabbage palm (*Sabal palmetto*) is also common. In more organic soils, cabbage palms can grow very tall, but on the limestone and scanty soil of the rocky pinelands, they seldom reach more than 15 feet and are regularly called palmettos rather than cabbage palms (their proper name). Both of these small palms are extremely resistant to fire.

The rocky pinelands also sustain numerous hardwood shrubs and small trees, including live oaks and nearly all of the tropical hardwood species that dominate the hammocks. Pinelands that have burned within the past three or four years are very open, with almost nothing blocking the view between the tops of the palmettos (usually about four to five feet) and the lower limbs of the pines, which may begin over 20 feet up the trunk. Pinelands that have not burned within five years begin to reveal the vision-blocking invasion of the hardwood species. If this invasion continues unchecked for several more years, a subsequent fire can be devastating, especially in particularly dry conditions. Not only are the hardwoods severely pruned or killed, but with so much fuel, the pines themselves may be killed, resulting in a barren landscape with dead pines standing like tombstones.

In the absence of fire for two to three decades, a pineland will take on the characteristics of a hammock in the process of succession. The dense shade produced by the encroaching hardwoods prevents seedling survival by all but certain shade-tolerant hammock species, such as pigeon plum. A "young" hammock can be recognized by the remaining presence of tall pines. Once established, the hammock itself becomes fire resistant because near ground level there is little fuel available to support a fire, except in extreme droughts, when the soil itself may burn. Thus, while pinelands are a fire-climax, ecologists consider the tropical hardwood hammocks to be the climax community of upland succession in southern Florida.

Endemic Plants and the Rocky Pinelands

Botanists are especially interested in the rocky pinelands of southern Florida. Other plant communities of the region are populated almost entirely by plants that occur in other areas of the world and have invaded the region only recently. The rocky pinelands, however, harbor many "endemic" species (found only in a certain region and nowhere else on earth). The rocky pinelands of the southern Everglades and nearby portions of the Atlantic Coastal Ridge harbor at least 20 endemic species* that are not found even in the pine flatwoods of the sandy, acid soils in the remainder of the state. Examples include a milk pea (*Galactia pinetorum*) and a narrow-leaved poinsettia (*Poinsettia pinetorum*).

The occurrence of so many endemics in this community indicates long-term continuity of conditions favorable to their survival and isolation. As a result, they have evolved into new, distinct species or varieties. Like slash pine, most are adapted to fire, not by physically withstanding the heat, but by other mechanisms

---

* The number of endemic plants varies considerably among sources (i.e., Craighead[39] states 100 "plants"). Much of the variation comes from whether only "species" or "taxa" are counted, the latter term including species, subspecies, and even named varieties. Also, recent discovery that many plants formerly considered endemic to southern Florida also occur in the Bahamas has reduced the list.[11]

that usually involve reproductive cycles. If the changes in southern Florida through the glacial cycles could be viewed, the perpetuity of the pinelands would probably be observed, while the tropical vegetation and freshwater wetland communities would come and go. This would lead to the conclusion that the really characteristic plant community of southern Florida, through time, is the pineland.*

---

* Fossil pollen and other records indicate that communities dominated by slash pine, how-ever, must have been restricted in distribution during glacial times when the peninsula was dominated by plant communities adapted to much drier conditions.[256]

7

Mangrove Swamps*

T hroughout the world, wherever coastlines are protected from the direct action of waves, the area between tides supports wetland vegetation. In temperate climates, this protected intertidal area is dominated by various grasses and related plants. For example, in northern Florida's Gulf and Atlantic coasts, where sufficient protection exists, the often extensive intertidal areas are salt marshes, dominated by smooth cordgrass (*Spartina alterniflora*) and black rush (*Juncus roemerianus*). While both of these species are found in southern Florida, they are restricted to much smaller areas by a completely different intertidal community: the mangrove swamp.

Types of Mangroves[236, 237]

Mangroves are tropical trees that are adapted to salt water and to the rigors of tides. While many species are called mangroves throughout the tropics of the world, only three species are normally referred to as mangroves in Florida: the red mangrove (*Rhizophora mangle*), the black mangrove (*Avicennia germinans*), and the white mangrove (*Laguncularia racemosa*). The common names of these species may be mystifying at first. The red mangrove derives its name from a red layer *under* the thin, gray bark. "Black" refers to the color of the bark of mature black mangrove. The white mangrove has the lightest colored bark of the three species, but it is not really white. These three "true" mangroves are ecologically related by their adaptations to a common habitat, but they are members of different plant families. Another species, the buttonwood (*Conocarpus erecta*), is sometimes called a mangrove. It belongs to the white mangrove family and is best suited to slightly higher ground around the edges of mangrove swamps. A "silver-leaved" variety of buttonwood is used as a native landscape plant in southern Florida.

---

* The primary citations for this chapter are References 95, 127, and 168.

FIGURE 7.1
Extensive root development of an old red mangrove in the interior of a mangrove swamp along the lower Oleta River, northeastern Dade County, Florida (January 1972). (Photo by T. Lodge.)

Red Mangrove

The red mangrove (Figure 7.1) is adapted to its intertidal habitat by a most curious feature, which has made this species a prime subject for photographers and landscape painters. It develops aerial "prop" roots from locations sometimes many feet high up on the trunk, or even from branches. These roots arch out and then down to the soil. The roots themselves often develop prop roots, sometimes with a half dozen successive arching tiers reaching outward from the tree.

The prop roots provide two adaptive functions. First, they form a massive support for the tree, with prop roots reaching in all directions. Red mangroves normally occupy the most exposed, outer locations of mangrove swamps, where waves and tidal currents are the most rigorous and where survival requires a firm grip. Second, the prop roots are an adaptation to hydric soils. Because intertidal soils are continuously saturated with water, there is no opportunity for atmospheric oxygen to get into the soil. Root growth and function require oxygen as do other plant tissues (as discussed earlier in the case of cypress trees in freshwater wetlands). Pores (white spots on the thick part of the exposed root just above the soil) exchange gasses with the atmosphere and supply oxygen to the subterranean root system. This adaptation also allows red mangroves to survive in the interior of mangrove swamps. Under certain conditions—usually relatively lower elevation—red mangroves may dominate the interior, where black mangroves usually rule.

Red mangroves do not require salt water. In fact, Hurricanes Donna (1960) and Betsy (1965) pushed huge numbers of seedlings inland into freshwater marshes in Everglades National Park. Many of these are thriving today and can be seen along the main park road between Mahogany Hammock and Paurotis Pond. A group is even growing among some cypress trees, which are highly intolerant of salt. Red mangroves, however, do much better in estuarine habitats, where sea water and fresh water mix.[39]

Black Mangrove

The black mangrove is also adapted to hydric soils, but its approach is opposite that of the red mangrove. Black mangrove roots beneath the soil send small extensions upward, reaching as much as a foot above the soil surface. They also function to exchange gases and bear the long name pneumatophore. Pneumatophores are functionally comparable to cypress knees, but are so numerous that the whole forest floor seems to be covered with them wherever black mangroves are found, mostly in the interior of mangrove swamps where tidal action is very sluggish, but shallow flooding may be prolonged.

White Mangrove, Buttonwood, and the Buttonwood Embankment

The white mangrove can occur almost anywhere in a mangrove swamp or protected shoreline, but it is most abundant in the higher elevations, such as the edges of mangrove swamps adjacent to uplands. Under stressed conditions, its roots may form short, branched pneumatophores and its lower trunk may produce small adventitious roots. Compared to the respective adaptations of the black and red mangroves, however, these growths are minor. White mangroves can colonize available upper intertidal areas rapidly, sometimes forming dense, nearly pure stands. A distinctive identifying character is two obvious glandular openings on the petiole (leaf stem) where it meets the leaf.

The buttonwood occupies still drier sites where tidal flooding is unlikely. It lacks the special seedling and root adaptations of the true mangroves, but it is tolerant of saline soils. Its sensitivity to frost restricts it to southern Florida, where it is primarily coastal in distribution, but occasionally is found inland in pinelands or at the edges of tropical hardwood hammocks. Buttonwoods form forests along the inland edges of mangrove swamps, saline coastal lakes, and tidal creeks. These forests are on naturally occurring levees of peat soil that stand about two feet above mean sea level, leading to the name *buttonwood embankment*. In many places, the buttonwood embankment acts as a dam, impounding shallow fresh water behind it during the rainy season and often creating an audible rush as high water spills over it. These areas are important to and rich in wildlife.[39]

Mangrove Swamps and Everglades Wildlife

The mangrove swamps of Everglades National Park cover more than 500 square miles, forming dense forests from the freshwater marshes seaward to the open waters of Florida Bay and the Gulf of Mexico (Figure 7.2). This area amounts to about two thirds of all the mangrove forests in Florida. The forest development is most spectacular along tidal rivers, including Roberts, North, Watson, Shark, Harney, Broad, and Lostmans rivers in the mangrove zone of the park. While mangroves visible to most people seldom exceed 50 feet in height, many specimens along these rivers reached nearly 100 feet before Hurricanes Donna and Andrew and severe cold fronts during the 1980s toppled and pruned them to a much lower level.

The transition from the freshwater marshes of the Everglades to the tidal mangrove swamps begins in a series of creeks, which reach like fingers from the mangroves into the marshes (Figure 7.3). The creeks, such as Rookery Branch, form the headwaters of the tidal rivers previously named. The rivers and their tributary creeks, coupled with the buttonwood embankment and its associated freshwater pools, are of great importance to wildlife. Freshwater flow from the marshes keeps the tributary creeks fresh to lightly brackish most of the time, with saline conditions reaching the creeks only during times of extended drought or storm-driven tides.

Alligators play a prominent role in the mangrove swamp tributary creeks; their activities add substantially to the ecological function of the area by allowing interplay between the upper tidal and freshwater marsh areas. Alligators keep the upper reaches of creeks open and provide numerous trails into the marsh; these trails are advantageous to aquatic life moving between the creeks and marsh.[39] The alligator's "enhancement" of the creeks provides wading birds with an increased availability of prey, much of which is juvenile fishes of estuarine and marine species.[137] This enhancement was likely a key reason why the historic wading bird rookeries of Everglades National Park were located in these areas (Figure 7.4).

Mangrove Swamps and Marine Fisheries

Mangrove swamps can be extremely productive. Given the right combination of tidal "flushing" (movement of water and suspended material in and out of the swamp) and freshwater runoff, mangrove swamps are among the most productive natural communities in the world. These conditions are present where the giant Shark River Slough and Big Cypress Swamp pour their fresh waters into the

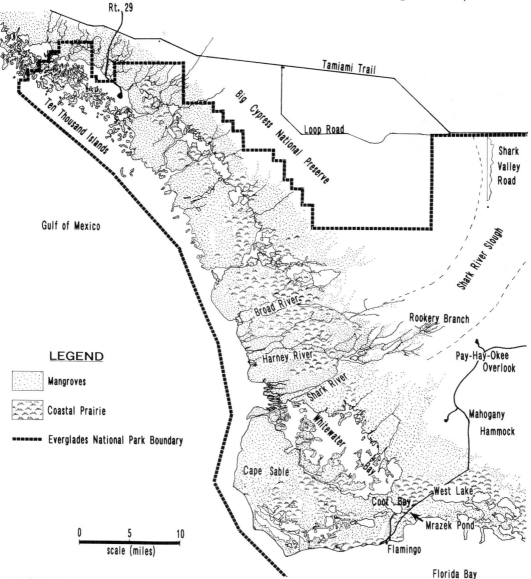

Rt. 29

Tamiami Trail

Big Cypress National Preserve

Loop Road

Shark
Valley
Road

Ten Thousand Islands

Gulf of Mexico

Shark River Slough

Broad River

Rookery Branch

LEGEND

 Mangroves

Coastal Prairie

Everglades National Park Boundary

Harney River

Shark River

Pay-Hay-Okee
Overlook

Mahogany
Hammock

Whitewater Bay

Cape Sable

West Lake

Cook Bay

Mrazek Pond

0   5   10
scale (miles)

Flamingo

Florida Bay

FIGURE 7.2
Map of the expanses of mangrove swamp and coastal prairie in Everglades National Park.

Gulf of Mexico, across the very gentle slope of the Florida plateau in the western half of Everglades National Park. The freshwater input dilutes the saltiness of marine waters, making life easier for the mangroves by requiring less effort to exclude or excrete salt. Therefore, more of their photosynthetic catch of sunlight energy can be directed toward growth. The tidal flushing also helps to move nutrients around, making them more available, and to transport products (dead leaves, twigs, etc.) away. The combined interaction of the tides and fresh water

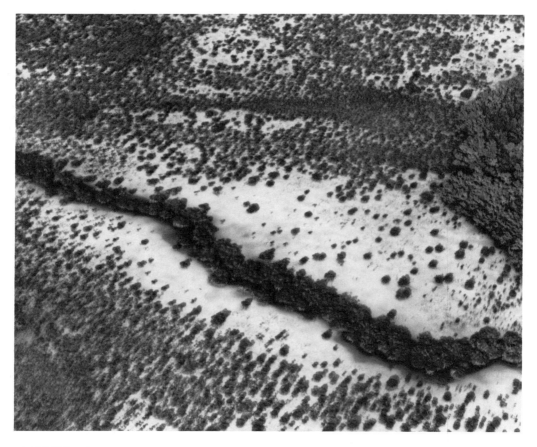

FIGURE 7.3
Aerial view of the inland invasion of young red mangroves along a small tidal creek in coastal marl prairie, Everglades National Park. The buttonwood embankment is at the right. (Photo by R. Hamer.)

can be compared to the work of a farmer in his field, tilling the soil, fertilizing, irrigating, and harvesting the final product. Just as the farmer's labor greatly increases the production of his land and the economy of the region, so too do the tides and freshwater runoff enhance the mangrove ecosystem and its benefactors.

The ecological values of the mangrove swamp arise from the fate of its products: the leaves, twigs, bark, pollen, and so forth. As these materials decay, they become food for marine life. Mangrove "detritus" (dead material) is eaten by many organisms at the base of an extensive food web, including shrimp, larval crabs, and small fish. In fact, the excellent fisheries of the Florida Keys, Florida Bay, and the eastern Gulf of Mexico owe much to the mangroves of Everglades National Park. Not only do the mangrove swamps provide a food source, but they and the associated tidal creeks and bays also act as a "nursery ground" for a great many marine species.

FIGURE 7.4
A mangrove used as a roost by wading birds stands defoliated in the great mangrove swamps of western Everglades National Park. (Photo by R. Hamer.)

Mangrove Swamps and Soil Building

Where they are protected from erosion, mangrove swamps are often soil builders. As individual trees die, their prolific root systems and varying loads of leaf and twig litter become peat. When sea level rises slowly, mangrove swamps are elevated by this soil-building process, at least until some major catastrophe (most often a hurricane) kills the trees (see Chapter 8). A common misconception is that mangrove swamps are land builders; rather, they merely keep pace with rising sea level, at least during times when sea level is rising slowly, as it has during the past 3000 years. Actual "land" building is probably a rare event that results when a hurricane deposits an unusual load of sediment into a mangrove swamp. However, some cases of land formation have been documented.[37, 250]

Mangroves and Hypersalinity

Red and black mangroves are well known for their means of surviving in high salinity which, because of evaporation, may greatly exceed that of normal sea water during dry periods. These two species have opposing adaptations to this condition. The red mangrove excludes salt from entering its roots. The black mangrove allows salt to enter, but then excretes it from its leaves. If it has not rained within the past day or so, the surfaces of the leaves will have a crust of salt. Thus, another identifying mark of the black mangrove is the taste test.[247]

Mangrove Reproduction and Dispersal

Dispersal of mangroves is by floating seeds, as might be expected for an intertidal plant. Floating seeds explain how mangroves got to southern Florida and explain how occasional mangroves take root in northern Florida, only to be killed by cold weather. Of the three species, the black mangrove is the most tolerant to cold and survives farther north, with some individuals found in the Mississippi delta.

The seeds of the red mangrove are a special case. Called propagules, they actually mature and begin growing while still on the tree. Each propagule develops an elongated body that becomes the trunk of the seedling. The lower, brown-colored end is pointed like a missile, and this end develops roots if it lodges in soil (enough so that the tip is in the dark). When the propagule is mature, it drops from the tree and may lodge in the soil directly, where it will begin growing. However, if it does not stick, it will float, on its side at first. After several days, it begins to change buoyancy and floats vertically. Finally, after several weeks, it sinks. If, by chance, during the course of its tidal travels it becomes lodged in a calm area shallow enough for growth, it rapidly sprouts roots from the bottom and an initial two leaves from the top.

Legal Protection of Mangroves

The story of mangroves and mangrove swamps does not end with a mere description of these interesting trees and how they live. Mangroves have been a focus of federal, state and county legislation in Florida, most of which emanated from studies that began in the late 1960s and pointed out the great ecological benefits that mangrove swamps provide man and wildlife alike. They also aid in shoreline stabilization, particularly during hurricanes (see Chapter 8). Mangroves take the initial rage of storm-tossed seas and thus protect man's interests further inland.*

Mangroves and Mosquitos

Mosquitos of the mangrove swamp are normally abundant and aggressive, and a smaller biting insect, the sand fly or "no-see-um" (technically a biting midge, or ceratopogonid) can be even worse. It is a popular misconception that mosquitos do not breed in salt water. The proper statement is that freshwater mosquitos do not breed in salt water. The saline waters of mangrove swamps throughout the tropics of the world, however, are breeding grounds for many species of mosquitos. In coastal Florida, the salt-marsh mosquito (*Aedes taeniorhynchus*) is abundant. This small species has distinctive black and white striped legs, abdomen, and proboscis. Winds occasionally blow hordes of them into residential areas. However, in a mangrove swamp during any warm month of the year, the combination of mosquitos and no-see-ums can defy description— like trying to explain the idea of infinity to a child! The mosquito "problem" is described in Chapter 5 on tropical hardwood hammocks.

Visiting a Mangrove Swamp

You can enter a mangrove swamp to see all this for yourself, but stepping over thigh-level prop roots and swatting mosquitos at the same time requires athletic ability, patience, and a touch of masochism. Some think a machete is required to get through a mangrove forest. My own experience is that using a machete results in blisters at the least, and possibly a major injury, in addition to falling far behind those who merely take the time to step over and walk around obstacles. The best way is to use the boardwalk at West Lake in Everglades National Park, with a fresh application of mosquito repellant, unless it is a cool, breezy day during the dry season, in which case repellant is not needed.

---

* The perceived value of mangroves is often confused by assuming that a mangrove tree itself is the value of the ecosystem. However, there have been many demonstrations that mangroves can even grow in lawns, if given care. It should be recognized that the value of a mangrove comes from its living in a viable tidal swamp or mangrove island where ecosystem "goods and services" are produced and exchanged. Outside of that context, a mangrove tree does not have any special environmental value.

8

Coastal Lowland Vegetation...and Hurricanes!

The awesome power of hurricanes to reshape nature—as well as man's existence—is of obvious importance in southern Florida. Their ecological impact can be seen in many plant communities, especially forested areas, but it is most prominent in the lower elevation, coastal communities, where a combination of wind and moving water rearranges both trees and terrain. Were it not for hurricanes, certain coastal habitats would not exist.

My own experience with hurricanes began on a normal, sunny summer morning, August 22, 1992. I had stopped at a camera shop at Suniland on South Dixie Highway, where the air was abuzz with concern. A hurricane had developed in the Atlantic and was then some 800 miles east of Miami and heading west. Its name was Andrew.

Inexperience with hurricanes shaped my indifference to the impending danger of Andrew. I had moved to Miami in 1966, and in the ensuing years, there had been several close calls but no hits. My timing had been lucky: Hurricane Betsy had passed the previous year and the ferocious Donna in 1960. In October 1966, Inez brushed South Florida but failed to come ashore. On September 2, 1979, David took direct aim at Miami, just like Andrew. My wife was into her ninth month of pregnancy, and despite our doctor's opinion, we worried about possibly having to bring our daughter into the world in the midst of a raging storm. Winds became alarming in the first morning hours of September 3, but Miami was spared as David veered north to Palm Beach and Stuart. With some justification, I thought Andrew would do the same. Little did I know that within 40 hours of my visit to the camera shop, in the last hours of darkness on Monday morning, August 24, my family and some friends would be prisoners in our house, witnessing the terrifying roar of hurricane winds.

By 7:00 a.m. that morning, the winds had dwindled enough, save some frightening gusts, for us to survey Andrew's damage. Our house suffered little, but trees and power lines in our neighborhood were in ruins. Most of the coconut palms that had lined both sides of the street lay on the ground, snapped off at the base and pointing southwest. We quickly determined that we had taken the northeast quadrant of the storm and assumed that we must have been very close to the eye to have incurred such damage. With no electricity and little means of getting reliable news, it took several days to understand how wrong we were. The northern edge of the eye had actually passed nearly ten miles south. Communities that had bordered it, from Kendall to Homestead and Florida City, lay in shambles. Over 80,000 homes in a 25-mile-wide path had been destroyed beyond practical repair.[87]

This personal experience is passed along to relay the message that individual people do not often endure hurricanes (the most powerful of tropical cyclonic storms, defined as having winds in excess of 74 miles per hour). From the perspective of natural history, however, their regular influence on coastal environments is an inescapable fact, worthy of considerable attention by ecologists.

Impacts of Hurricane Andrew on the Everglades[44, 53, 57a, 132, 199]

The full impact of Hurricane Andrew on the natural history of southern Florida will not be known for several years from this writing. Small but intense, Andrew followed a westward track through the heart of the mainland area of Everglades National Park (Figure 8.1). Its path of heavy impact was about 30 miles wide, where estimates put maximum sustained winds at 150 miles per hour, with gusts to 200. Initial surveys revealed massive defoliation of coastal mangrove swamps, 70,000 acres described as "knocked down," and extensive damage to virtually all tall hammock trees in its path, most of which were either stripped of leaves and branches or completely toppled. Within a few weeks following the hurricane, vines began to overtake the skeletons of the remaining hammock trees, smothering the slow recovery of damaged but living trees. Early impressions indicated that less apparent destruction occurred in the second-growth pinelands, attesting to the great strength of the younger slash pine. Older, taller pines were commonly broken, but invariably high above the ground, the grip of their roots refusing to yield. However, within a year, on the order of 90 percent of the pines in Andrew's path had died, having succumbed to pine bark beetles and weevils because of their weakened condition.[49] The least damage was observed in the vegetation of the freshwater marshes, but there was apparent extensive loss of resident aquatic life, such as mosquitofish.[133]

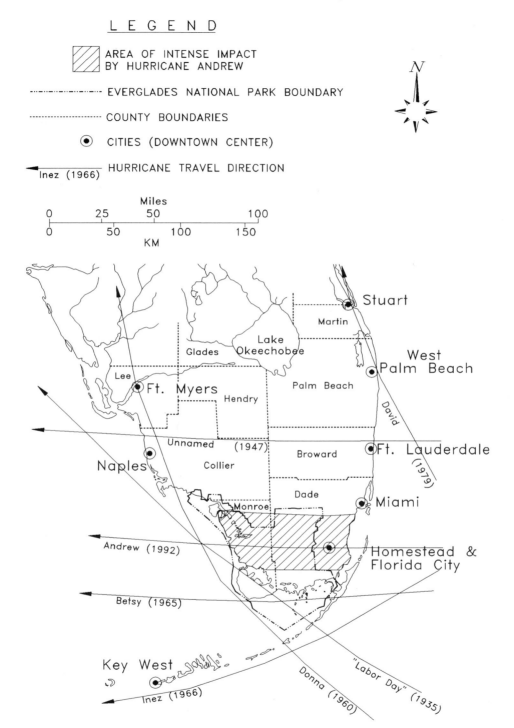

LEGEND

⬛ AREA OF INTENSE IMPACT
BY HURRICANE ANDREW

▬▬▬▬▬ EVERGLADES NATIONAL PARK BOUNDARY

----------- COUNTY BOUNDARIES

◉ CITIES (DOWNTOWN CENTER)

◄───── HURRICANE TRAVEL DIRECTION
Inez (1966)

N

Miles

0   25   50   100
├───┼────┼──────────┤
0   50   100   150
KM

FIGURE 8.1
Map of the tracks of selected hurricanes affecting southern Florida from 1935 to 1992.

Hurricane Frequency and Environmental Impact in Southern Florida*, [76]

Was Andrew an unusual hurricane? For the people who experienced its fury, the answer is emphatically, "yes!" During the prior 60 years, however, there were two major hurricanes that wrought destruction specifically to the area of Everglades National Park: the relatively small but intense Labor Day hurricane of 1935, which Andrew resembled, and the much larger Hurricane Donna of 1960. These two storms altered the ecology of the coastal portions of the park (established in 1947) to such an extent that hurricanes are an unavoidable topic in the ecology of southern Florida—as basic as alligators.

Based on almost a century of observations, southern Florida is struck by a hurricane on an average frequency of once every three years. The coastal portions of Everglades National Park have been struck about once every ten years, and "great hurricanes" (with winds of 125 mph or more) strike the park about four times per century. The peak hurricane season is mid-August through most of October.

Destruction by hurricanes is due to one or more of four factors: wind, waves, storm surge, and rain. For most people, storm surge is the least understood factor, and close to the coast it is potentially the most damaging aspect of a storm. Storm surges are a combination of the "mound" of water (the storm tide) induced by the hurricane's low barometric pressure and the rough seas and current driven by the storm's winds. Storm surges of severe hurricanes often reach more than ten feet above regular tidal levels. The relatively small, but deadly, Labor Day hurricane of September 1935 swamped Matecumbe Key in the upper Florida Keys with an 18-foot surge, and Andrew's surge neared 17 feet.[222]

Impacts of Hurricane Donna[39]

Until the full impact of Andrew is measured, the best documentation of the effects of a hurricane on the Everglades ecosystem lies in the studies that followed Hurricane Donna. On September 10, 1960, Donna's eye crossed the middle Florida Keys with sustained winds of 140 mph and a storm surge of about ten feet. From the Keys, it headed for the shoreline of Everglades National Park with a track that kept the northeastern quarter of the storm (usually the most powerful part) over the mangrove swamps and other coastal environments, as the eye moved to Naples and Fort Myers on Florida's southwest coast.

---

* The reader should be aware that numerous hurricanes and tropical storms affected the Everglades during the period discussed in the text, in addition to the examples cited. These included a giant September 1947 hurricane that added to already extensive flooding in its westward passage across the central Everglades, with a path of heavy impact about 100 miles wide.

Some 120 square miles of mangrove swamp was devastated. It is estimated that half of the trees were killed. Many were sheared off by the high winds, and those left standing were stripped of most branches and leaves. These effects were noted immediately after the storm, but then a peculiar sequence of events began. Within a few weeks, as expected, intact trees began recovering, with new growth evident everywhere. Soon, however, the process stopped, leaving a complete kill of mangrove trees over extensive areas. Why did the recovery cease? The answer lay in the impact of the storm surge. The ten-foot-deep mass of moving water forced inland by Donna carried with it a tremendous load of lime mud (marl) from Florida Bay and the bays and lakes inside the coastline. The mangrove forests slowed the flow of the water, causing much of this load of marl to be deposited as a blanket up to about six inches deep over the forest floor. Examination proved that this was enough to interfere with the critical oxygen supply required by the tree roots, killing virtually all of the trees in low-lying areas. Many trees on slightly higher elevations survived and, oddly enough, even a large number of mangroves that had been completely uprooted, but with blocks of soil still clinging to the roots, also survived. An adequate oxygen supply continued to be available for the roots of these displaced trees.

Donna had another major effect. Prior to Donna, the mangrove forests had supported epiphytic plants in great abundance and variety, including air plants, orchids, and ferns. In the damaged area, it was estimated that 90 percent of these perished, either due to outright removal by the high winds or subsequent exposure to direct sunlight in the dead, leafless forests. Perhaps a full century will be required for this profusion of epiphytes to become reestablished in similar abundance in the affected mangrove swamp areas. It is highly possible that some very rare, perhaps recently established species were completely exterminated from Florida. Thus, while some hurricanes may be due credit for bringing the seeds of tropical plants, such as epiphytes, to tropical southern Florida (see Chapter 10), others may destroy established populations.

The Hurricane Legacy: Coastal Lowland Vegetation[39]

The patchwork of other kinds of lowland habitats in and around the great mangrove swamps is primarily the result of hurricanes. Even some areas of higher elevation that today support species other than mangroves are the result of deposits that killed a previous forest. These non-mangrove coastal lowlands support interesting and often beautiful plant communities. If one species had to be identified as the most ubiquitous of these salty habitats, it would be the buttonwood (*Conocarpus erecta*). Buttonwoods may stand as scattered individuals or may be closely spaced, forming a buttonwood forest. Many dead snags that look like aged driftwood are buttonwood trunks, while others are mangrove "tombstones" left from Donna or even from the Labor Day hurricane of 1935 (Figure 8.2). The variability of these saline areas is great, and ecologists have named numerous plant associations including black rush marshes, saltwort

FIGURE 8.2
Dead mangrove and buttonwood snags are evidence of a former mangrove swamp decimated by Hurricane Donna (1960) and now a coastal prairie dominated by saltwort and glasswort. A stiff-leaved wild pine sits prominently on the roots of a toppled and long-dead mangrove near Flamingo, Everglades National Park. (Photo by R. Hamer.)

marshes, coastal marl prairies, and scrub mangroves. Many such habitats are present near the town of Flamingo, at the end of the main road in Everglades National Park.*

The most picturesque of these areas is the saltwort marsh, named for the abundant light, yellow-green ground cover of saltwort (*Batis maritima*). Prior to Hurricane Donna, few areas were dominated by this shade-intolerant species, but Donna opened extensive habitat and spread its seeds. This decorative little

---

* Many of the "higher" coastal areas (1.5 to 2 feet above sea level), especially near Flamingo, previously supported coastal hammocks, dominated by cabbage palms and salt-tolerant tropical hardwoods such as West Indian mahogany, buttonwood, and Jamaica dogwood (*Piscidia piscipula*). Cleared for lumber and charcoal, these areas are now coastal prairies that were not born of hurricanes.

plant has a woody stem that arches over and supports numerous short branches with fleshy, salty-tasting, inch-long leaves shaped like miniature cucumbers. Saltwort has a maximum height of three feet, and flats dominated by it are very open. With care not to trip over the stems, and with the decided advantage of long legs, you can hike through these areas. The soil is normally of marl and is firm but usually wet, except in the height of the dry season.

In addition to buttonwood and saltwort, other plants in the areas opened by hurricanes (or sometimes by man) include sea daisy (*Borrichia frutescens*), glasswort (*Salicornia virginica*), sea purslane (*Sesuvium portulacastrum*), and scattered black, white, and red mangroves. Dead snags and live buttonwoods support numerous epiphytes, making these open, sunny areas ideal for observation and photography. The epiphytes include several kinds of air plants (*Tillandsia* spp.) and some orchids, the most common of which is the butterfly orchid (*Encyclia tampensis*) (Figure 8.3). However, when exploring any habitats near the coast, be prepared for mosquitos in all but cool, breezy, winter weather.

FIGURE 8.3
An osprey consumes a fish on a mangrove snag remaining from Hurricane Donna and shared by a blooming butterfly orchid at lower left, near Flamingo, Everglades National Park. (Photo by R. Hamer.)

Hurricanes and Glacial Cycles

Hurricanes, and their progenitors—tropical storms and tropical depressions—are the result of warm sea surface temperatures and proper atmospheric conditions. During winter and spring months, these conditions do not exist in the tropical North Atlantic, Caribbean, or Gulf of Mexico. Beginning in June and extending into November each year, the conditions are ripe to spawn and sustain hurricanes. Because hurricanes require warm, open water, they cannot develop over land the way that tornados do. Extrapolate these requirements back to the times of the interglacial sea that existed prior to 100,000 years ago. The warmth of those times and the expanded sea surface (the Gulf of Mexico was significantly larger then, in part at the expense of Florida) must have provided conditions that spawned hurricanes in great numbers. Thousands of hurricanes have probably revised southern Florida's coastal environments during the warmer parts of each sea level excursion through the glacial cycles.

9

Coastal Estuarine and Marine Waters[*]

T he coastal receiving waters of the Everglades watershed are Florida Bay to the south and the Gulf of Mexico to the southwest (Figure 9.1). However, the interconnecting labyrinth of bays and tidal creeks lying within the coastal outline confuse the land/sea border. Furthermore, the great expanses of mangrove swamp grading inland into freshwater marsh make the distinction between land and water even more difficult. Excellent fishing and bird watching are available in these waters, but an experienced guide is usually required for exploration.

Florida Bay: A Geologist's Classroom

In addition to its obvious interest to fishermen, Florida Bay (Figure 9.2) is famous among geologists. Processes going on within Florida Bay have answered geologists' questions concerning the formation of certain fine-grained limestones. Most limestones have an obvious origin because they are made of identifiable fossils. In the limestones of southern Florida, there are corals in some and bryozoans in others, which makes their origins fairly obvious. Some limestones, however, contain very few fossils and instead consist of very fine grains of calcium carbonate that do not have an obvious origin. These fine grains may also form the bulk of other limestones, as a filler between fossils.

Studies conducted in Florida Bay have resolved this question of limestone origin. The puzzle in Florida Bay that attracted scientific interest was the vast amount of lime mud (marine marl) on the bottom. Grains of this material are

---

[*] The primary citation for this chapter is Reference 100.

FIGURE 9.1
Northeastern Florida Bay. This view looks westward over Little Blackwater Sound (left foreground) and Long Sound (center), Everglades National Park, May 1972. (Photo by T. Lodge.)

similar to those of the fine-grained limestones and are far smaller than those in the underlying limestone floor (the Miami Limestone formation) which is composed mostly of oolites. The oolites (small egg-shaped grains) are known to have had a physical-chemical origin and were deposited in a marine environment before the last glaciation, when sea level was substantially higher than today. Thus, the youngest layers of the limestone floor date back about 100,000 years, to the time when sea level started to recede during the last glacial cycle. The marine marl layer averages about six feet thick, yet it has all accumulated in only about the last 4000 years, as determined by radiocarbon dating. This is the time when

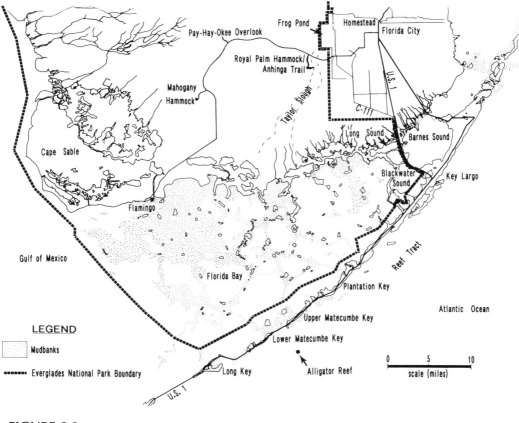

FIGURE 9.2
Map of Florida Bay.

rising sea level first flooded Florida Bay with marine waters. Previously, as evidenced by deeper sediments lying on the rock floor, the freshwater Everglades extended over the area of today's Florida Bay.

The search for the origin of Florida Bay's lime mud did not have to go far. The bottom is almost completely covered with tropical marine vegetation, some of which is turtle grass (*Thalassia testudinum*), a "higher" plant with true roots, leaves, and flowers (although they are too small to be noticed). The bulk of the vegetation, however, is many species of algae. Numerous among these are species that have a root-like "holdfast" and other parts that superficially resemble stalks and leaves of flowering plants. These types of algae have stiffened bodies, which aid in maintaining upright posture. Species of *Penicillus* (the four-inch-tall "shaving brush" alga), *Halimeda, Udotea,* and others are abundant in the area. The stiffening in these species results from calcium carbonate particles produced within their tissues, and these particles appear identical to those that compose the bulk of the sediments. Furthermore, when these short-lived algae die, they disintegrate so completely that all that is left is their deposit of calcium carbonate particles. This process is comparable to the production of marl by periphyton in freshwater marshes (see Chapter 3).

Studies on the rates of growth of these bay-bottom algae have confirmed that they could easily be responsible for a large proportion of the mud that has accumulated during the past 4000 years. In some locations, the mud is ten feet deep. The nature of the sediments, however, is much more complex than just fine-grained carbonate material. The sediments are reworked by worms and other burrowing organisms, to make pellets of carbonate. The shells of tiny foraminifera, bryozoans, and molluscs become incorporated, and physical sedimentation adds precipitated carbonates.[19, 252] Given the proper circumstances, such as another reduction in sea level, thus exposing this lime mud and other carbonate sediments to atmospheric weathering, they would consolidate and recrystallize into a rock layer, much of which would be fine-grained limestone.

Another puzzle in Florida Bay is the configuration of the bottom, which is a patchwork of deeper areas, called "lakes," separated by shallow mudbanks. Mangrove islands, which are important for roosting and nesting birds, dot some of the these mudbanks. The lakes are about four to six feet deep, while the mudbanks are generally exposed at low tide, where they are used extensively by a great variety of fish-eating birds, including the great white heron, reddish egret, spoonbill, and occasionally the native American flamingo. The lakes and mudbanks make navigation a considerable challenge, and low tide can leave an inexperienced fisherman stranded many miles from land.[61, 188, 252]

The topography of the bottom of Florida Bay appears related to the shapes of the myriad of lakes and bays that lie inside the coastline. A few of these, such as West Lake, Coot Bay, and the much larger Whitewater Bay, are easily accessible to Everglades National Park visitors by boardwalk or boat tour. It is theorized that as sea level rises, the bays and lakes are gradually enlarged, principally by hurricanes which drive the waters so fast that they erode the bottoms and shorelines of the bodies of open water and deposit much of the material in the surrounding mangrove swamps. The mangrove swamps are finally overtaken by rising sea level, leaving mudbanks in their place with a few scattered mangrove islands, until they too are overtaken (Figure 9.3).

In turn, mangroves invade further inland, establishing first along new tidal channels of the once freshwater sloughs and along alligator trails. New open-water bodies are formed, again by storms, from the remaining marshes between the advancing mangrove swamps. Marsh vegetation cannot withstand water movement driven by hurricane winds and are "hollowed out," forming shallow open-water bodies that become tidal lakes and bays as sea level continues to rise. This historical sequence makes a modern excursion from the freshwater Everglades marshes through the mangrove swamps, the tidal creeks and bays, and into Florida Bay equivalent to a trip through time in Florida Bay.

Evidence of this sequence lies in buried sediments, which verify that the freshwater Everglades once extended far out into Florida Bay. The lower portion of sediment core samples taken throughout the bay reveals the presence of peat that contains recognizable remains of freshwater Everglades plants. Above this layer, many core samples contain peat derived from the roots of mangroves, but

FIGURE 9.3
A mangrove island stands in the background while young red mangroves find a perilous, storm-threatened existence in shallows dominated by turtle grass and algae in Florida Bay near Flamingo, Everglades National Park. (Photo by R. Hamer.)

it is apparent that Florida Bay never had extensive mangrove swamps like those found in the western portion of the park today. When Florida Bay first flooded with salt water, about 5000 years ago, sea level was rising a little too fast: about a foot per century. Any mangroves that could establish were soon overtaken by open marine waters, where marine sediments were laid down. Thus, the extensive upper sediment layers (mostly marl) are the marine deposits. As sea level rose, a forced succession of habitats occurred, from freshwater marsh, to a scattering of mangrove swamp areas, and finally to the open, shallow marine habitat that has remained for some 4000 years.[43]

The configuration of Florida Bay, with its pattern of mudbanks and shallow lakes and its location pinched between the Keys and the mainland, restricts circulation, thus making the bay prone to extremes of salinity, temperature, and reconfiguration by storms, especially hurricanes. Hypersalinity has received the most attention from ecologists. The main avenue of freshwater flow from the Everglades is, and historically was, the Shark River Slough, which releases its water to the west of Cape Sable into the Gulf of Mexico. Only the relatively small eastern area of the Everglades has provided fresh water directly to Florida Bay, principally through Taylor Slough. Thus, in dry seasons, evaporation may exceed freshwater input, and salts are concentrated. While historic data are few, this condition undoubtedly recurred naturally but has been aggravated by hydro-logic modifications of the Everglades (also see Chapter 18).[155, 225, 252]

The Gulf of Mexico

The westernmost coastal waters of Everglades National Park and tropical southern Florida lie in the Gulf of Mexico. The line between Florida Bay and the Gulf of Mexico is indefinite, but the interesting features of the bottom of Florida Bay end on an approximate northwest-southeast line between Cape Sable (the southernmost prominence of dry land of the Florida mainland) and Long Key in the chain of Florida Keys.

The shoreline of the Gulf of Mexico, west of Cape Sable, looks very different than that of Florida Bay. Along Florida Bay, much of the shoreline is a low ridge of dry land, at least during normal tidal cycles. Northwest of Cape Sable, how-ever, mangrove swamps and a labyrinth of tidal rivers are the normal shoreline feature. At the westernmost part of Everglades National Park, near the town of Everglades City, the apparent contiguous shoreline of mangrove swamps is preceded by a myriad of small mangrove islands. This region is called the Ten Thousand Islands.

Throughout this area, currents along the shore of the Gulf of Mexico have carried silica sand southward. This type of sand is familiar on the beaches of the Atlantic and Gulf coastal states. Here it mixes with the carbonate mud and abundant shell fragments to make a firmer bottom than the carbonate mud of Florida Bay. The combination of the firmer bottom and the freshwater outflows from the Everglades and Big Cypress Swamp make an ideal habitat for the oyster, a species that has a major ecological impact on shallow tidal habitats wherever it is abundant.

Oysters and Mangrove Swamps[129]

Oysters are related to clams, and like most clams, they are "filter feeders." Oysters strain water through a fine network in their gills, trapping minute particles including bacteria, plankton, and non-living material, all of which are the oyster's food. Unlike free-moving types of clams, oysters grow attached to a surface and are not free to move around once their planktonic larval form has settled down. They do best in areas where tidal currents transport food to them, remove wastes from them, and prevent the accumulation of smothering sedi-ments. However, strong currents make it difficult for the attachment of larval oysters because of shifting sediments. Thus, a good site for a larval oyster's attachment is the shell of an existing, solidly anchored oyster shell, whether dead or alive.

Oysters live best in brackish water and at depths where they are periodically exposed by low tide. These conditions are unfavorable for some major oyster predators, including snails called oyster drills and stone crabs. The result of all these factors is that oysters tend to form large colonies in shallow tidal areas in the mouths of creeks and rivers. Because the colonies grow fastest where the current is swift, they tend to elongate across the path of tidal currents, forming long structures called oyster bars which tend to block tidal channels. As might be imagined, an oyster bar can be a rude awakening for a naive boatman! The tough, four-inch shells "cemented" together can cut through a hull and have ruined countless props.

In southern Florida, oyster bars present conditions conducive to the establishment of mangrove seedlings, especially the red mangrove. Under favorable conditions, a mangrove forest can overtake an oyster bar. This encroachment, together with sedimentation processes along the Gulf of Mexico shoreline of Everglades National Park, has led to an interesting phenomenon: it appears that mangrove swamps have actually been extending seaward in portions of this area, in spite of slowly rising sea level. Longshore currents have supplied the sand and shells that add to any local carbonate production, making shallower water. Oysters became established, and then mangroves took over. This process has helped to create one of the world's largest mangrove swamps, much of which is located in the western part of Everglades National Park. Although sea level has been increasing over the past 3000 years, its slow pace has allowed oyster bars and mangroves to prosper.

Part III

Biogeography of Southern Florida

10

Origins of
the Flora and Fauna*

Before mankind emerged as a travel agent for a myriad of species ranging from rats and carp to Brazilian peppers and Australian pines, flora and fauna were dependent on natural means of dispersal for colonizing new regions. Biologically, South Florida is a "new land" because of its history of alternating submergence and emergence and because of its major changes in climate, both of which have eradicated many species of plants and animals through the glacial cycles. The important factors that determined the region's "new" flora and fauna were its geographic setting at the edge of the tropics, from which it is isolated by water, and its land connection by a long peninsula to temperate North America. The flora and fauna present when Columbus arrived in the New World had arrived in the region by their own modes of dispersal** and had succeeded based on their tolerance of the climate, the habitats, competition, and numerous other factors.

Tropical versus Subtropical[39, 57a, 236]

Whether or not southern Florida is truly tropical is often debated. One way that biologists define *tropical* is by climatic conditions that exclude freezing

---

* The primary citations for this chapter are References 123, 138, 163, and 256.

** Native Americans inhabited central Florida at least 14,000 years ago and southern Florida 10,000 years ago and more recently began travel to and from Cuba.[28, 32, 156] While it is possible that Native Americans aided in the dispersal of tropical plants and animals to southern Florida, it is doubtful that they had a significant impact because they did not use the vast majority of tropical species; therefore, they would have had no reason to carry these species in their travels.[27]

temperatures on land or that rarely include temperatures below 68°F in aquatic or marine environments. Based on these criteria, the offshore marine environment of the Florida Keys is tropical, but the land and the shallow, inshore waters of southern Florida are not. Cold fronts can blast through southern Florida and plunge temperatures from the mid-70°F one day to freezing the next. Notable freezes that killed many tropical plants occurred in 1894, 1895, 1899, 1917, 1962, 1977, and 1989 (personal observation of the author), and damaging frosts recur about every other year.[230, 245]

Despite occasional damaging freezes, a great many tropical plants exist naturally in southern Florida, which suggests a more useful definition of "tropical": climatic conditions conducive to the survival of plants that are damaged by freezing temperatures. By this definition, the extreme South Florida mainland and the Florida Keys are tropical. The number of tropical species that occur there, and that do not occur only slightly farther north on the peninsula, is impressive.

The climate for tropical plants is significantly better near the coast, especially the east coast and the Florida Keys, where the Florida Current and coastal marine waters contribute protective warmth. It is this maritime influence that has extended the climate for tropical plants north of the Tropic of Cancer, which is normally the defined northward limit of the tropics. Inland, or coastally only a few miles north of the latitude of Miami, tropical plants are commonly killed by cold weather or pruned to a miniature stature; it is therefore useful to distinguish the climate of those areas as *subtropical* based on vegetation.

The warm maritime effect is demonstrated in the distribution of native tropical trees. The geiger tree (*Cordia sebestena*), with its beautiful orange flowers, and the manchineel (*Hippomane mancinella*), renowned for its poisonous fruit, thrive in the Florida Keys but are restricted in Miami, where they survive only in protected locations. Less sensitive tropical trees such as the gumbo limbo (*Bursera simaruba*), the strangler fig (*Ficus aurea*), and the West Indian mahogany (*Swietenia mahagoni*) are abundant at the latitude of Miami, but become restricted northward to the immediate coast. Some tropical species reach as far north as Cape Canaveral on Florida's east coast and Tampa Bay on the west coast. The sensitivity of tropical plants to cold is the main factor that determines their northern limits, and a northward restriction to the coast is the standard pattern of their distribution.

Many authors have shunned the term "tropical" to describe the southern tip of Florida, in favor of "subtropical." The latter is used by climatologists to describe much of the southern United States. To label the entire region "subtropical" misses what the vegetation clearly demonstrates: tropical plants "know" the difference compared with the lands farther north or in the interior of the region.

Elements of the Flora[138, 183]

During the relatively recent glaciation, when the Florida peninsula was cooler and drier than today, a temperate flora thrived. The tropical flora has only invaded the region in the last 5000 years. The warming trend following the last glacial period had apparently progressed enough by then that the climate resembled that of today, and a tropical flora began to flourish.

Speaking only of *vascular plants* (those with roots and stems, as opposed to algae and fungi), about 1650 species now exist south of the latitude of Lake Okeechobee. Just over 60 percent of these are allied to the tropics, with about 56 percent occurring elsewhere in the Caribbean region. The remaining 40 percent of southern Florida's vascular flora is composed of temperate species derived mostly from the coastal plain of the southeastern states. Within this overall flora, there is a significant number of endemic plants; that is, species found only in Florida, which have evolved there due to isolation (see Chapter 6). They amount to about 9 percent of the flora of the region, and their distant origin is a mix of temperate and tropical sources.*

Another interesting statistic is that 65 percent of the flora is herbaceous and 35 percent is woody types, namely trees and shrubs. Among the herbaceous species, about half are tropical, but among the woody species about 77 percent are tropical.[138] Thus, the strongest tropical heritage resides among the woody species. Within these species, this heritage is most pronounced in trees, quantified as follows:

> Of the approximately 130 tree species native to South Florida, 112 (87.5 percent) have an otherwise tropical distribution; in Florida they are at the northern limit of their distribution. The remaining 18 species (12.5 per cent) represent temperate trees with a wider distribution in the eastern United States; they are therefore at the southern limit of their distribution in South Florida.[236]

Origin of the Tropical Flora

How did tropical plants get to southern Florida? Three modes of dispersal account for the distribution of most plants: their seeds or spores are transported by wind, water, or animals. Dispersal by animals across water is usually by birds.

Trees

Almost all of the tropical trees that occur naturally in southern Florida are also found in the West Indies.* Most species occur in hardwood hammocks, and a few occur in mangrove swamps (four species including buttonwood) and freshwater wetlands (two species). Seeds of the hammock species are commonly carried by birds; thus, migrating birds, which eat seeds while in the tropics and

---

* The percentages presented here are from Long.[138] They include a significant number of introduced "naturalized" plants (species introduced by man and reproducing in the wild, outside of cultivation), together with the native flora.

then release them in droppings in Florida, probably account for the arrival of many tropical trees. The white-crowned pigeon, for example, is a prolific disperser of tropical seeds throughout the West Indies and southern Florida.[198] West Indian mahogany, however, is an example of a wind-dispersed species.

The mangroves, as well as many other coastal wetland plants, have floating seeds and obviously were dispersed to Florida via ocean currents. The two tropical trees of freshwater wetlands—pond apple (*Annona glabra*) and cocoplum (*Chrysobalanus icaco*)—are an interesting biogeographic problem. Both occur in tropical West Africa and tropical America, including the West Indies. Both have fruits that are eaten by animals, and birds probably account for their dispersal to Florida, perhaps from Cuba. Nevertheless, their occurrence on both sides of the Atlantic is puzzling.[175, 198, 236]

Epiphytes

Another prominent feature of the tropical vegetation of the region is the abundance of plants that grow upon other plants: the epiphytes. Although some parasitic plants live in the region, epiphytes are not parasitic by definition and are wind-dispersed. The major epiphytes of the region are bromeliads, orchids, and ferns.

Bromeliads

South Florida's most obvious epiphytes, the bromeliads (often called "air plants"), are relatives of the pineapple. All native bromeliads are normally epiphytic, but they may grow on the ground if they happen to fall. Due to its widespread occurrence in the South, the best-known species is Spanish moss (*Tillandsia usneoides*), which has an unusual growth form. Most other species resemble the top, leafy part of a pineapple; however, close inspection of Spanish moss reveals that it is a "string" of small plants attached by a stem. Spanish moss is not tropical and is more common farther north in Florida, with its range extending as far as north as Virginia. Most of Florida's 15 other bromeliads are tropical, are restricted to southern Florida, and their seeds are distributed by the wind. A spectacular and common species in the Everglades region is the large stiff-leaved wild pine (*Tillandsia fasciculata*). Its leaves may exceed three feet long (although usually half that) and its small flowers are enclosed by conspicuous, attractive bracts that are usually bright red but may be variable in color. It blooms from January through the summer.[36]

---

* It is interesting to note that dispersal of trees has essentially been one way. Very few temperate trees are found in the tropical islands of the West Indies, although fossil pollen studies indicate that a larger representation occurred in the distant past. The limited occurrence of the live oak (*Quercus virginiana*) in western Cuba is one of the few examples that exists today and probably represents a population that remains from a time when the climate was more favorable for temperate plants there.[236]

Orchids

Less obvious epiphytes, unless they are in bloom, are the orchids. There are about 20,000 species of orchids worldwide; they occur in both tropical and temperate climates. By far the greatest diversity of species occurs in the tropics, where most are epiphytic; temperate orchids are mostly terrestrial. Florida harbors about 100 types of orchids; temperate, mostly terrestrial species are predominant in northern Florida and tropical species (many of which are epiphytes) are predominant in southern Florida. The most common epiphytic orchid in the region is the beautiful butterfly orchid (*Encyclia tampensis*), a tropical species that blooms primarily in the late spring and early summer.

Orchids are best known for their spectacular flowers. The vegetative parts of many species are rather small and often go unnoticed. Intricate flowers usually indicate specialized means of pollination, and the degree to which many orchids have evolved in their means of pollination is amazing. This specialization, which usually employs specific insect pollinators, is most developed in tropical species and has undoubtedly prevented many from becoming established in southern Florida. Because distribution of the tiny orchid seeds is by wind, they are able to reach South Florida from the nearby islands of the West Indies, germinate, and grow into mature orchids. The problem is pollination: such specialized species cannot reproduce without the necessary types of pollinators. As a result, most of the tropical orchids of the region are types that are normally capable of self-pollination. While southern Florida does have numerous tropical orchids (about 60 species are shared with Cuba), many more could have become established if the proper pollinators were present.[36, 141]

Ferns

Numerous types of ferns also grow as epiphytes, particularly in very damp locations. Of about 100 species of ferns found in Florida, almost 60 are tropical and are limited to southern Florida. In contrast to bromeliads and orchids, which generally occupy fairly well-lit sites, ferns abound even in the dense forest shade; they grow on trees with rough bark, dead logs, or rocky walls of solution holes (see Chapter 5). The golden polypody or serpent fern (*Phlebodium aureum*), a common tropical species in the Everglades region and Central Florida, normally grows on the trunks of cabbage palms. Ferns reproduce by spores, which are technically not seeds but serve the same purpose and are wind-blown.[36, 119]

Marine Flora

All of southern Florida's marine flora reached the region from the West Indies on ocean currents. Although many species of tropical algae occur, only five species of marine "grasses" are found, the most prominent of which is turtle grass (*Thalassia testudinum*). The brief pulses of cold air that invade the Florida peninsula in winter have little effect on the marine flora. In fact, tropical marine species extend much farther north than the tropical land vegetation, some reaching as far as the coast of North Carolina.[139, 227]

Hurricanes and Dispersal

The violent, destructive impacts of hurricanes were discussed in Chapter 8, but hurricanes (and tropical storms) also assist in the inland transport of floating seeds on storm surges and probably serve to transport airborne seeds. The cyclonic motion makes any hurricane that enters the Straits of Florida a potential carrier of airborne seeds from Cuba and the Bahamas to southern Florida. Hurricane Inez (October 1966) was a good example. The eye crossed Cuba from the Caribbean into the straits and moved northeastward to a stopping point in the Bahamas just east of Miami. Of minimal hurricane strength at that point, it reversed direction and followed the Florida Keys into the Gulf of Mexico. Its particular track was almost ideal for carrying airborne seeds from Cuba and the Bahamas to Florida.

Proximity and Dispersal

The success of wind-blown dispersal in particular is dependent upon the proximity of the land masses. It is evident that the proximity of Florida to the West Indies during the past 5000 years has allowed the influx of tropical plants. Before that time, southern Florida was too cold to have supported sensitive tropical plants. During earlier, warm interglacial periods, the southern tip of the Florida peninsula may have been so far north that dispersal of tropical plants to Florida from the West Indies may have been more restricted.[236] During those times, the continued existence of many tropical plants that had become established in Florida would have been more dependent on their local reproduction. "Re-seeding" would have been limited.

Modern sea level and climate may be special, transient conditions for Florida. The southern tip of the state is warm enough to support the invasion of a sensitive tropical flora that was not present until the last 5000 years.[138] If global warming continues, the shoreline of Florida will continue to retreat northward, farther out of the reach of natural colonization by many tropical plants. Based on cyclic changes involving the earth's orbit and rotation, another glaciation should be starting now,[184] but it is apparent that the polar ice caps are melting, sea level is rising, and man's activities are aggravating this process. Thus, we will probably never know if the natural glacial remission was complete.

Origin of the Temperate Flora[138, 139, 236]

While much interest in the Everglades region is focused on tropical vegetation, the plants that originated in temperate North America also lend character to the ecology of the area. They are particularly important in freshwater wetland communities of the region, including the lower elevation pinelands, wetland tree islands, and marshes.

Trees

The temperate trees of southern Florida dispersed southward into the region by a variety of means. Bald cypress (*Taxodium distichum*) has floating seeds, which enable it to spread through freshwater wetlands and streams. Red maple (*Acer rubrum*) and coastal-plain willow (*Salix caroliniana*) have wind-blown seeds. Other species, such as live oak (*Quercus virginiana*) and probably sweet bay (*Magnolia virginica*) and red bay (*Persea borbonia*), are dispersed by mammals and birds.[128]

Marsh Vegetation

The seeds of most marsh plants are dispersed both by wind and by floating on water. The plants of the Everglades itself—the freshwater marshes—are mostly temperate. However, many of the marsh plants do not have a pronounced tropical or temperate affinity. Sawgrass (*Cladium jamaicense*), the most abundant plant in the Everglades, occurs from Virginia to Florida as well as in tropical areas of Central America and on the islands of the West Indies. Marsh plants, which consist mostly of grasses and similar herbaceous (non-woody) species, are adapted to regular adverse conditions that may kill their top portions. Cold weather, drought, and fire are periodic rigors of the interior freshwater marshes from which the vegetation normally recovers by regrowth from the roots, which are protected in the soil. This ability to "escape" adverse conditions allows a great many herbaceous plants to be distributed across the tropical/temperate boundaries that are so important to woody and epiphytic plants. Thus, southern Florida has an intriguing mix of native temperate flora that has been naturally "invaded" in recent time by a great many tropical species due to the unique and transitory circumstances of its geography and climate.

Origins of the Fauna[163, 256]

The mobility of animal life makes it impractical to define species as "tropical" or "temperate" based on their tolerance to cold weather. The actual freezing point is less important for animals than it is for most plants, because many can avoid exposure to adverse temperatures by seeking some form of shelter. In this respect, animals are similar to the marsh plants discussed earlier. Nevertheless, patterns of distribution of individual species, or of groups of related species such as genera or families, make it possible to determine most origins.

The distributions and relationships of the native animals in southern Florida show that the terrestrial and freshwater species are almost completely of temperate, North American origin. However, fossil evidence shows that Florida has been invaded by South American land animals several times over the past 2.5 million years, since the land bridge through Central America came into existence, only to perish during times of cooler, drier climate. In today's native fauna, forms that can disperse through the air (flying or drifting) show a mixture of origins.

Some are from the tropics, but most are of temperate origin, except for the isolated islands of the Florida Keys where, among spiders for example, tropical species predominate. Only the marine animal life of the region has a strong overall, natural tropical heritage.

Beginning with invertebrates and continuing with fishes, amphibians, reptiles, mammals, and birds, origins of South Florida's fauna, as well as the roles played by these animals in the ecosystem, are revealed in the following chapters. While the biogeographic approach is useful for organizing knowledge, it is the ecological functions performed by animal life that are critical to the performance of the Everglades ecosystem.

11

Invertebrates

Invertebrates do not have a backbone and include all animals except fishes, amphibians, reptiles, mammals, and birds. While non-biologists may think that these exceptions rule out most forms of animal life, such is not the case. Some familiar invertebrate groups are sponges, jellyfish, corals, worms, clams, snails, octopuses and squids, spiders, scorpions, insects, crabs and lobsters, shrimp, starfish, and sea urchins. This list only scratches the surface by naming the well-known groups, leaving out such creatures as rotifers, bryozoans, copepods, thousand-leggers, and arrow worms, among numerous others. Invertebrates include such a wide variety of animal life that all conceivable modes of dispersal apply to invertebrates of one kind or another.

Marine Invertebrates[35, 108]

Most marine invertebrates have planktonic larval forms that are dispersed by ocean currents. A tropical starfish need not crawl from Cuba to Florida's tropical waters through the depths of the Straits of Florida. Such migration would be impossible for an adult, but the larval form can make the trip passively. In fact, this type of dispersal accounts for the prevalence of tropical marine invertebrates in the waters surrounding southern Florida, including the famous reef-building corals. At the same time, larval dispersal greatly complicates studies of the effects of overfishing on invertebrates, such as the highly valued Florida spiny lobster (*Panulirus argus*), which is the tropical epicurean equivalent of the Maine lobster. Effective regulation of fishing requires an understanding of larval movements. The spiny lobster population of Florida Bay is established by larvae that arrive from the ocean, but the location of the parental populations has not yet been established (i.e., Florida waters or even elsewhere in the Gulf/Caribbean area).[45]

Not all southern Florida marine invertebrates are strictly tropical. Two commercially important examples are the blue crab (*Callinectes sapidus*) and the pink shrimp (*Penaeus duorarum*). Blue crabs are found from Cape Cod all the way southward to Florida and the Gulf of Mexico. Pink shrimp are found as far north as Chesapeake Bay, but are much more common in the warmer waters from the Carolinas southward.[88] Florida Bay and the mangrove swamps in Everglades National Park are extremely important nursery grounds for the huge population of pink shrimp harvested near the Dry Tortugas, a group of islands west of Key West.

Freshwater* Invertebrates

Most freshwater invertebrates of southern Florida are not from the tropics, but rather from the fresh waters of North America. The Everglades does not have a great diversity of freshwater invertebrates due to its limited type of habitat and its nearly tropical climate, which many temperate species cannot tolerate.

Florida Applesnail**

The Florida applesnail (*Pomacea paludosa*) (Figure 11.1), a tropical species, is an important freshwater mollusk in the Everglades. This large, globose, dark brown snail, whose adults measure over two inches in diameter, is found in Florida's wetlands, lakes, and rivers. It is restricted to the warm water of spring runs in the northern part of the state, although it probably reached Florida during an earlier, warmer time from northern Mexico, where a closely related species (a probable ancestor) occurs.[231]

The applesnail is peculiar in its dual ability to extract oxygen from water using gills or from air using the equivalent of a lung. To breathe air, it crawls to a point near the surface, extends a tube called a siphon, pumps air in and out of a chamber, and then returns to its underwater browsing on algae. Its masses of

---

* The term *fresh water* is found as both a single word and as two words (or hyphenated in older literature). Accepted practice is now two words for the noun phrase and one word for the adjective; thus freshwater organisms live in fresh water. However, there is frequently lack of uniformity, and even Florida's wildlife agency is named the Florida Game and Fresh Water Fish Commission.

 Fresh water is defined as having salinity (saltiness) low enough that there is no appreciable sea water mixed into it. Sea water contains about 3.5 percent salt by weight. Ecologists usually express salinity in parts per thousand (ppt); thus, sea water is about 35 ppt. The upper limit of salinity normally used to define fresh water is 0.3 ppt.[117] Between these values, water is termed brackish, typical of estuaries and coastal wetlands. It is important to recognize that this term has a specialized meaning, and it should not be confused with meanings that could be assumed from the two individual words, namely that "fresh water" might be water that is not "stale" or "polluted." Fresh water is—unfortunately—often polluted!

** Nomenclature is standardized here to follow Turgeon et al.;[240] however, the name usually appears as "apple snail" (two words) in the Everglades literature.

FIGURE 11.1
Florida applesnail and its eggs, laid on a pickerel weed leaf. (Photo by T. Lodge.)

white eggs deposited on plant stems above the water line are common in the Everglades during the summer. Despite their appearance, the eggs are not tolerant of drought. They hatch in less than a month and must enter water immediately. Adults can survive by aestivating while buried in algal mat or mud, but most perish if flooding does not recur within about four months. Thus, applesnails are most abundant in or near permanent water.[193]

The Florida applesnail is an important topic in the Everglades, due to its predation by a variety of wildlife including young alligators and numerous birds. The most publicized example is the snail kite, a hawk-like bird that feeds exclusively on the applesnail and is thus completely dependent upon water levels that maintain the snail's habitat. The limpkin (Figure 11.2), a wading bird which is related to cranes, is also heavily dependent on the applesnail.[111]

Seminole Rams-horn

Much smaller than the Florida applesnail, the Seminole rams-horn (*Planorbella duryi*) is widely found throughout the Everglades. Its shell is a flattened coil that typically reaches about three quarters of an inch in diameter. The species is endemic to the Florida peninsula where it is confined to fresh water, and its shells are a good marker for soils developed in fresh water.*, [231]

---

* Helisoma marl[39] refers to this species. The genus *Helisoma* is closely related to *Planorbella*, and the name is often improperly substituted for *Planorbella*.

FIGURE 11.2
A limpkin dines on a Florida applesnail. (Photo by R. Hamer.)

The Seminole rams-horn, like the Florida applesnail, grazes on algae. In turn (and together with young applesnails), it is an important prey of the redear sunfish, a snail "specialist." This snail can survive for an undetermined time by burrowing in moist soil. Because it can survive in very limited water in underground limestone cavities, it often is found in short-hydroperiod marshes (especially where solution holes retain water) as well as in more permanently flooded habitats.[92, 136] Furthermore, it can disperse rapidly during high water by "riding" on the underside of the surface. This ability, best exhibited by smaller individuals, involves attachment of the snail's foot to the water surface. Wind-blown currents can then propel the snails hundreds of times farther than their own slow crawling would allow. Other snails, including juvenile Florida applesnails, also travel by this means, and large numbers of snails are sometimes observed along the windward sides of tree islands or sawgrass patches in the Everglades, long distances from their apparent sources. It is important, however, that the water be deeper than just a few inches in order for this mode of transport to be effective.[193]

Crayfish

About 300 species of crayfish, which look like miniature facsimiles of the marine Maine lobster, occur in the fresh waters of North America. Florida hosts

FIGURE 11.3
A two-inch crayfish, probably overcome by recent fire, is testimony to subterranean life during the dry season in a sparse sawgrass marsh near Pa-Hay-Okee Overlook, Everglades National Park. (Photo by R. Hamer.)

50 species, but only four occur in the southern half of the peninsula and only one, *Procambarus alleni*, lives in the Everglades.* Like most crayfishes, *P. alleni* has no common name, but it should be called the Everglades crayfish (Figure 11.3). It is endemic to the Florida peninsula and is adapted to the motionless waters of marshes and to the alternating wet and dry seasons of the region. It lives in underground burrows during the dry season and browses on algae and small invertebrates over the marsh bottom during the wet season. Females carry eggs while still underground at the end of the dry season, and young crayfish populate the newly flooded marshes at the inception of the rainy season. Crayfish are important prey for largemouth bass, pig frogs, young alliga-

---

* Another species, *Procambarus fallax*, is found north of the Everglades primarily in moving-water habitats, including the freshwater streams of the Loxahatchee River system, which historically drained from the northeastern edge of the Everglades.[98]

tors, and wading birds, particularly the white ibis.[111] Because of its reproductive timing, *P. alleni* is one of the first abundant prey in the Everglades early in the wet season.[72, 98, 99, 179, 192]

Riverine Grass Shrimp*

Another important crustacean in the Everglades food chain is the riverine grass shrimp (*Palaemonetes paludosus*). The genus *Palaemonetes* has many species, several of which inhabit the estuarine and marine habitats of the region, but *P. paludosus* is the only abundant species in the freshwater habitats of the region. It resembles the edible pink shrimp, except that it grows only to about three quarters of an inch and is so transparent that it is almost invisible. Unlike the crayfish, it cannot survive a complete drydown, and therefore its populations are low in short-hydroperiod wetlands, except near refuges such as alligator holes. In more permanently flooded habitats, such as sloughs, however, its abundance can exceed 100 individuals per square yard. Riverine grass shrimp feed on algae and, in turn, are eaten by virtually all predatory fishes. However, except for the white ibis and little blue heron, which catch significant numbers, most wading birds do not feed on them because of their small size and cryptic guise; yet through the food chain, the riverine grass shrimp is a highly important link to all fish-eating birds in freshwater habitats.[60, 135]

Aquatic Insects

Aquatic insects are also important in the food chain of the Everglades, as they are in all freshwater habitats. Numerous insects (e.g., water scavenger beetles, water boatmen, and giant water bugs) are totally aquatic but have retained the ability to fly and will attempt to relocate if their environments dry up. Most species that inhabit aquatic habitats, however, live in the water only as larvae and emerge to live in the air as adults. Prominent examples are mosquitos, mayflies, damselflies, and dragonflies. Their dual lifestyles represent a significant transfer of energy from aquatic to terrestrial environments, where flying insects are important in the diets of many birds.[180]

Dragonflies (also called darning needles, mosquito hawks, and snake doctors) represent highly visible and interesting examples of insects associated with aquatic habitats. Their larvae are predatory and feed on anything they can catch and overpower, including other aquatic insects (and sometimes even their own kind), worms, snails, and occasionally even small fish and polliwogs. In turn, they are eaten by larger fishes and certain wading birds, notably the little blue heron and the white ibis. Adult dragonflies feed on insects; they splendidly consume large numbers of mosquitos but also eat larger insects

---

* Nomenclature is standardized here following Williams et al.;[257] however, the name usually appears as freshwater prawn in the Everglades literature.

including beetles, wasps, and other dragonflies. The eastern pondhawk* (*Erythemis simplicicollis*), which is two inches in size, is a voracious example that is sometimes cannibalistic.

Of the 86 species of dragonfly that reside in the Florida peninsula, only about a dozen reproduce in the Everglades, primarily because the limited type of habitat (marsh) is not suitable for those species that require the moving water of streams in their larval stages or forested habitat as adults. The Everglades species are mainly temperate, and common examples are the Halloween pennant (*Celithemis eponina*); the powerful green darner (*Anax junius*), which is unusual in that it is a migratory species; the four-spotted pennant (*Brachymesia gravida*); and the eastern pondhawk. Within all the habitats of southern Florida, about half of the dragonflies found are tropical and half are temperate. Because they are strong fliers, long distances do not limit most adult dragonflies in colonizing new areas, at least for those species that do not require shade. The important factor that does limit their distribution is suitability of the aquatic habitat for their larvae.[57b, 178]

Terrestrial Invertebrates

Spiders

Although they are terrestrial invertebrates, many spiders utilize a means of dispersal that is more normally associated with plants, namely wind. Young spiders of many species are known to climb to a high point on a plant or obstacle and release a length of silk into the air. When the trailing silk is long enough, a breeze can pull the spider into the air. The process is called ballooning, and it accounts for the occurrence of many tropical spiders in southern Florida. Most of the species in the Florida Keys are of tropical, West Indian origin and probably arrived there by ballooning. The southern mainland hosts a mixture of tropical and temperate North American spiders.[125, 142]

Insects

Dispersal of tropical insects to southern Florida has accounted for the presence of many of the species. Flying insects are frequently carried by storms, and a few actually migrate from location to location. However, even though the wind may assist tropical species in their dispersal to southern Florida, other factors determine whether or not they can become established. Cases in point are demonstrated by several butterflies.

---

* Florida's dragonflies have been assigned English names, as distinct from common names. Because few people distinguish between types of dragonflies (or most invertebrates for that matter), specific common names have not evolved as they have, for example, for birds and fishes.[57b]

Butterflies

The zebra butterfly (*Heliconius charitonius*) is of tropical origin, but it occurs widely throughout southern Florida and reaches northern Florida during summer, where it may overwinter in mild years.[24] This handsome, slow-flying species has black wings with bold yellow stripes. Its distinctive elongated wings give it a much greater wingspan than is typical for its size among North America butterflies, although this feature is common among related tropical butterflies of Central and South America. Its occurrence in Florida is based on the presence of several species of tropical passion flower vines. Butterflies as a group depend upon specific kinds of food plants for their larval development, and the caterpillars of the zebra butterfly feed on passion flower leaves. The close relationship of the butterfly to this food plant goes beyond nourishment. Passion flowers contain a bitter substance that the caterpillars accumulate without harm. The caterpillars, and the butterflies in turn, are then unpalatable to potential predators such as birds and lizards, which reject them after their first taste. The relatively tough butterflies often show the wear of an attempted taste or two, with older specimens often having frayed wings.[104, 219] The biogeographic principle is that some species cannot relocate without the presence of another species. In this case, the zebra butterfly could not have successfully invaded Florida without the previous establishment of its food plant.

Several other tropical butterflies are present in southern Florida, due to the presence of certain tropical plants. In a few cases, the rarity of the food plant accounts for very limited distribution of the respective butterfly. Examples are the endangered Schaus' swallowtail (*Heraclides aristodemus*), which feeds only on the tropical hammock tree, torchwood (*Amyris elemifera*); and the rare Florida atala (*Eumaeus atala*), which feeds on the Florida coontie (*Zamia pumila*), a type of cycad that superficially resembles a fern. Although previously abundant, the Florida coontie is now highly restricted in distribution due to land development, and thus the butterfly has also become rare.[71]

Florida Tree Snail

The Florida tree snail (*Liguus fasciatus*), a tropical species, is a terrestrial mollusk of great interest in Florida. It is restricted to the extreme southern portion of Florida, but it also occurs in Cuba. It feeds on small epiphytes, such as lichens, on tropical hammock trees, usually those with smooth bark. These large snails grow to a length of about two-and-a-half inches and are highly variable in coloration, which is the basis for the high level of interest among shell collectors and other observers. As many as 50 distinct varieties have been found, ranging in color from pure white, to various banding and streaking patterns of green, yellow, pink, and brown, to individuals almost completely colored in brown and yellow tones.[71]

The Florida tree snail undoubtedly originated in the tropics, but how it invaded Florida is a mystery. Perhaps the explanation lies in its ability to aestivate through several months of the dry season by cementing its shell closed against the bark of a tree. It is possible, for example, that it could travel on a large tree limb that broke off in Cuba and floated to southern Florida with its branches above the water. The exact circumstance that allows a tree snail to land successfully by this rafting mode requires a vivid imagination!

Importance of Invertebrates*

Because of their lower position in the food chain, invertebrates are of great importance in the diets of many Everglades predators, including fishes, young alligators and crocodiles, wading birds, otters, and many others. Except for commercially important species such as the pink shrimp, however, relatively little is known about their ecological roles. For example, the importance of the applesnail to wildlife is widely acknowledged, but in fact the species has been little studied.[176, 232] The lack of even a summary publication on Everglades invertebrates—most needed for aquatic species—presented a substantial challenge in writing this chapter. Information is lacking on innumerable species, which affords ample opportunity to future students of the Everglades ecosystem.

---

* Subsequent to the first printing of this book, a study of Everglades invertebrates was published. See Rader, Russell B. 1994. Macroinvertebrates of the northern Everglades: species composition and trophic structure. *Florida Scientist* 57(1,2):22–33.

12

Freshwater Fishes*

I first visited Everglades National Park in September 1966, during the height of the annual rainy season. Out of curiosity, I stopped at one point to observe water flowing through a culvert under the road to Pa-Hay-Okee Overlook. The area was teeming with fishes, all easily visible in the clear, shallow water. I was fascinated watching gar, bass, several kinds of sunfish, and innumerable smaller species all jockey about in the sunlight. The sight was a delight to a budding ichthyologist from Ohio who was beginning graduate school in Florida.

Freshwater fishes are a mainstay of the Everglades food chains. They provide the diet for alligators, otters, wading birds, and other predators. The historic Everglades succeeded as a wildlife exhibition because its annual cycles promoted the growth and availability of a large freshwater fish biomass (the total weight of living matter in a given volume or area of environment, e.g., pounds per acre of fish). The more common or noteworthy species that occur in fresh waters of the Everglades are listed in Table 12.1.

Primary Freshwater Fishes

Among freshwater fishes, there is a wide range in tolerance to salinity. Species that are intolerant of salty water are unable to stray into estuaries and are termed *primary freshwater fishes* by ecologists. Their dependence on fresh water restricts their ability to invade and colonize new areas, limiting their travels to freshwater routes. Where two river systems have no freshwater connection, primary freshwater fishes cannot move between the systems without help, such

---

* The primary citations for this chapter are References 51, 134a, and 137.

TABLE 12.1
Common or Noteworthy Fishes of the Fresh Waters of the Everglades

Common name[a]	Ecological group[b]	Scientific name[a]
Florida gar	2	*Lepisosteus platyrhincus*
bowfin (mudfish)	1	*Amia calva*
tarpon	3	*Megalops atlanticus*
American eel	3	*Anguilla rostrata*
golden shiner	1	*Notemigonus crysoleucas*
lake chubsucker	1	*Erimyzon sucetta*
yellow bullhead (butter cat)	1	*Ameiurus natalis*
walking catfish	1-i	*Clarias batrachus*
sheepshead minnow	3	*Cyprinodon variegatus*
golden topminnow	2	*Fundulus chrysotus*
marsh killifish	3	*Fundulus confluentus*
flagfish	2	*Jordanella floridae*
bluefin killifish	2	*Lucania goodei*
eastern mosquitofish	2	*Gambusia holbrooki*
least killifish	2	*Heterandria formosa*
sailfin molly	3	*Poecilia latipinna*
brook silverside	3	*Labidesthes sicculus*
common snook	3	*Centropomus undecimalis*
Everglades pygmy sunfish	1	*Elassoma evergladei*
bluespotted sunfish	1	*Enneacanthus gloriosus*
warmouth	1	*Lepomis gulosus*
bluegill (bream)	1	*Lepomis macrochirus*
dollar sunfish	1	*Lepomis marginatus*
redear sunfish (shellcracker)	1	*Lepomis microlophus*
spotted sunfish (stump-knocker)	1	*Lepomis punctatus*
largemouth bass	1	*Micropterus salmoides*
oscar	2-i	*Astronotus ocellatus*
Mayan cichlid	3-i	*Cichlasoma urophthalmus*
blue tilapia	2-i	*Tilapia aurea*
spotted tilapia	2-i	*Tilapia mariae*
striped mullet	3	*Mugil cephalus*

[a] Common and scientific names, arranged in phylogenetic order, follow Robbins et al.,[203] except for those in parenthesis which are vernaculars commonly used in Florida.[26] It should be noted that the name bream is used here to designate the bluegill, although it is often used for any similarly shaped sunfish, such as the redear sunfish, and is therefore equivalent to the term pan fish.

[b] Ecological group refers to salinity tolerance as described in the text: 1 = primary freshwater fish; 2 = secondary freshwater fish; 3 = peripheral freshwater fish. "i" indicates introduced, exotic species.

as a flood, a geologic event that causes a stream to change course, or the hand of man.

All of the native primary freshwater fishes of southern Florida arrived from temperate North America after the emergence of the region from the sea. None is from the American tropics. Largemouth bass, bluegills, redear sunfish, golden shiners, and yellow bullheads are common in the Everglades. These species occur widely in the United States, even as far north as the Great Lakes. On the other hand, relatively few of the many primary freshwater fishes in the southeast have been able to "invade" the Florida peninsula, despite convenient river and interconnecting wetland corridors. Only 20 species, which are tolerant of warm, slowly moving waters, have reached southern Florida naturally. Among them, several are distinctive Florida varieties, often recognizable by color patterns that differ from their northern counterparts.[20, 81]

The Florida largemouth bass, found only in peninsular Florida, is the most famous of the primary freshwater fishes of the region. It is a subspecies of the largemouth bass, which occurs naturally in North America from the Great Lakes southward to the Gulf and lower Atlantic coasts and is now widely stocked around the world. The largemouth bass is a spectacular game fish, and the Florida subspecies is especially renowned. It frequently grows to 10 pounds and occasionally exceeds 20, averaging substantially larger that the northern subspecies. Because it is more wary, the Florida subspecies is also more difficult to catch. Research has indicated that this behavior may have a genetic basis. The "catch per unit effort" is lower for Florida bass than for the northern subspecies, even when the two are raised in identical, side-by-side ponds.[262] Perhaps this innate wariness is an adaptation to the prevalence of large predators—alligators, herons, and egrets—in Florida's extensive wetland habitats.

Secondary Freshwater Fishes

Freshwater fishes with more tolerance to salt water and the ability to live there temporarily are termed *secondary freshwater fishes*. They can colonize new areas by swimming from the mouth of one river system into the next through interconnecting estuaries, salty coastal wetlands, or even sea water. This ability allows them to colonize faster than primary species.

Most of the eight species of secondary freshwater fishes in southern Florida are the killifishes and their relatives. Generally small and colorful, many of these fishes make good aquarium specimens. They originated in the American tropics and long ago spread into North America, where they evolved separately and produced many North American species. Although their distant ancestors lived in the tropics, the southern Florida species came from temperate North America. Examples include the inch-long eastern mosquitofish (the most abundant and ubiquitous fish in Florida), the least killifish (North America's smallest fish), the golden topminnow, and the flagfish (named for its beautiful pattern of red stripes which resembles the U.S. flag).

Peripheral Freshwater Fishes

The last group, the *peripheral freshwater fishes*, may live part or most of their lives in marine waters, but regularly move or stray into fresh water. While most species in this group remain near salt water, some can be found anywhere in southern Florida's freshwater habitats. The sailfin molly, a popular aquarium fish, is such an example; however, it is most abundant in brackish, coastal waters. Other examples of the more than 50 southern Florida species in this group include the American eel, tarpon, snook, several killifishes (other than those categorized as secondary), needlefishes, mullet, several gobies and sleepers, and occasionally even bull sharks (mostly juveniles).

Many peripheral freshwater fishes are tropical species, intolerant to cold weather. Tarpon and snook are often killed when confined to shallow waters that cool quickly during winter cold fronts, and many a sad fisherman has seen a trophy-sized snook floating dead following a severe cold spell.

The Florida Gar

The strangest fish in the Everglades is the Florida gar, whose bizarre appearance includes sharp, needle-like teeth that fill a long snout. Young gar have numerous dark spots and patches on an olive to yellow, long slender body. They darken with age so that adults appear mostly dark brown, especially when seen from above. Several types of gar exist in eastern and central North America, some of which are extremely large. The aptly named alligator gar (*Lepisosteus spatula*) is occasionally mistaken for an alligator and occurs from the lower Mississippi drainage basin to the rivers of the western panhandle of Florida.[80] Only the relatively small Florida gar, seldom longer than two feet, lives in the Everglades. (The much larger longnose gar (*Lepisosteus osseus*) historically only occurred north of the Everglades region, but has occasionally been found in canals in the Water Conservation Areas of the Everglades.) As with all gars, the Florida gar is predatory and is adept at catching smaller fish from schools by using a fast sideways snap. It is also capable of catching individual prey, pursuing them along the bottom or in dense tangles of vegetation. Using a slow, stealthy approach, this technique is effective on fish and grass shrimp.[106]

Florida gar are sometimes seen in huge numbers, which is the result of low water confining individuals from the expanses of Everglades marshes into limited aquatic habitat during the dry season. At these times, gar become prey for the alligator. The sight of a gar held in an alligator's jaws is a vision of prehistoric imagery (see Figure 12.2). In fact, gar have changed little from ancestors that dominated the earth's waters when the dinosaurs flourished; they even have primitive, interlocking scales that differ greatly from most fish. They also have the dual ability to breathe air and water and can be observed regularly rising to the surface of the water to renew the air in their swim bladders. The Florida gar should not be confused with the similarly shaped but unrelated needlefishes (*Strongylura* spp.), which are marine but commonly enter fresh water. Needlefishes are greenish, bluish, or silvery and have a translucent appearance, in marked contrast to the darker and opaque Florida gar.

Introduced Fishes

The Everglades now hosts several firmly established exotic species, most of which have "escaped" from the tropical fish trade. It was in the mid-1970s that I first saw one within Everglades National Park: a walking catfish in the jaws of an alligator at the Anhinga Trail. Today, introduced kinds are prevalent there, and the blue tilapia, oscar, and Mayan cichlid are among the most visible. Recently, no less than five introduced species were seen at the Anhinga Trail, not including the walking catfish, which usually goes unseen—except by herons and egrets. Since the late 1970s, many walking catfish have been observed in the beaks of these wading birds.

Freshwater Fishes and the Food Chain

Food chain relationships of the Everglades fishes are important in understanding the freshwater ecology of the region (Figure 12.1). Herbivorous species include the golden shiner, sailfin molly, least killifish, and flagfish. Other fishes, such as golden topminnows, marsh killifish, mosquitofish, bullheads, redear sunfish, and bluegills, feed on invertebrates, many of which graze on plants. Adult Florida gar, warmouth, and largemouth bass feed principally on other fishes. In turn, all of these fishes are the vital food supply for other predators (Figures 12.2 to 12.4).[106]

The important role that fishes play in the food chain can only occur where fishes are able to maintain populations, which is a problem in a region where aquatic habitats dry seasonally. The single species known to survive a complete drydown is the marsh killifish.* Actually, adults perish, but the buried eggs survive even during a lengthy drydown and hatch rapidly when flooding returns. Other species are only able to repopulate the Everglades marshes from an aquatic refuge, such as an alligator hole or a cavity in the rock. Mosquitofish are particularly adept at survival in very limited space, possibly even using crayfish burrows, so that they seem to come from nowhere upon the return of

---

* Observations[51] indicate that the bowfin may aestivate in algal mat or mud upon drying in the Everglades, which may explain why individuals sometimes appear far from permanent water soon after reflooding. However, because the species probably cannot withstand much desiccation, it may only survive a short drydown.

FIGURE 12.1
Common forage fishes of Everglades marsh habitats: sailfin molly (top), an adult male least killifish (upper center), golden topminnow (left center), flagfish (lower left), and the ubiquitous eastern mosquitofish. The larger specimens are about an inch-and-a-half long. (Photo by T. Lodge.)

FIGURE 12.2
Predator takes predator: an alligator eats a Florida gar. (Photo by R. Hamer.)

FIGURE 12.3
Predator takes predator: an anhinga swallows a warmouth, a species of sunfish that commonly invades shallow marsh habitat. (Photo by R. Hamer.)

summer rains. For small species such as the mosquitofish, only a few inches of water may suffice for their dispersal through the marshes from an aquatic refuge. Larger species such as the golden shiner, sunfishes, and Florida gar probably require closer to a foot of water depth, especially for spawning. However, little is known about depth requirements for marsh utilization by freshwater fishes, and this topic is in need of further research. It is important for numerous species to have unrestricted access to the open expanses of marsh with the return of each wet season, because freshwater fish are the main prey of most wading birds and numerous other wildlife species in the Everglades. Freshwater fish populations are the key ingredient in the overall success of Everglades wildlife.[47a]

FIGURE 12.4
Predator takes predator: a great egret eats a young largemouth bass. (Photo by R. Hamer.)

The Fisherman's Perspective

For a freshwater fisherman experienced in the eastern or central United States, the general impression in southern Florida is familiarity. Most of the freshwater species found in the interior are familiar, except perhaps the gar and now several introduced, exotic species. The surprise is in the peripheral species: it is not unusual to be fishing for bluegills and see a three-foot tarpon swim past!

13

Marine and Estuarine Fishes*

W hile the true Everglades is a freshwater habitat, its waters drain into and nourish the prolific estuarine and marine portions of Everglades National Park, the boundaries of which extend far into Florida Bay and the Gulf of Mexico. Marine and estuarine fishes in these waters command abundant interest from aquarists, connoisseurs, and fishermen, but they are also an important food supply for many fish-eating birds. This supply does not form an annually concentrated biomass like the freshwater Everglades, but it is more consistently available on daily tidal cycles.[188, 233]

Diversity of Marine and Estuarine Fishes

A vast number of marine and estuarine fishes occur in southern Florida. If the Florida Current and its deeper "counter-current" are included together with the nearby portions of the Gulf of Mexico and habitats of the various bays and estuaries around the mainland, the species number about 1000. If only nearshore waters are included, the number is well over 500, which is far greater than any other vertebrate group of southern Florida.

The richness of the marine fauna results from geographic setting and habitat diversity. The geographic setting in the edge of the tropics brings a commingling of tropical and temperate species to the waters of southern Florida. Tropical marine fishes have no problem reaching the area. In contrast to freshwater fishes, most marine species have minute planktonic larvae that are dispersed by ocean currents, and the Florida Current has been the highway for the arrival of tropical

---

* The primary citations for this chapter are References 133, 201, and 202.

marine fishes in southern Florida. If freshwater species began life as drifting plankton, many would be carried out to sea to perish. Adaptation has enabled most freshwater fishes to swim soon after hatching from eggs that stick to the bottom or to vegetation.

South Florida is abundant in habitat diversity, including:

- Deep, open ocean waters of the Florida Current in the Straits of Florida
- Coral reefs along the Florida Keys
- Continental shelf waters, which are narrow on the east coast but very broad in the Gulf of Mexico
- Estuaries and mangrove swamps
- Shallow-water grass flats of bays and coastal waters
- Protected, shallow, coastal embayments

The number of species of fish that occur in each of these habitats differs considerably. The marine waters within Everglades National Park harbor about 260 species of fish and include the last four habitats listed. Diversity is lower in these inshore waters, mainly because the tropical reef species do not find acceptable habitat there and substantially fewer species can tolerate the rigors of changing temperature and salinity that characterize these waters. The richest habitat is the coral reefs, where a prominent tropical ichthyofauna is present. A study of Alligator Reef, located east of the park, revealed 517 species.[225]

Game Fishes[225, 233]

Species of interest to Everglades National Park sport fishermen are listed in Table 13.1. Biogeographically, these game species include both temperate and tropical types. However, several tropical species, such as tarpon and great barracuda, travel far north into temperate waters, especially in the summer. Tarpon and snook regularly enter fresh water. The gray snapper, known locally as the mangrove snapper, represents the mostly tropical snapper family, but this particular species is extremely widespread in inshore waters from southern Brazil to New York, including the entire Gulf of Mexico and the Caribbean. The jewfish is a giant tropical grouper found in the tidal channels of Florida Bay. This species, now protected because of depletion by overfishing, can reach a length of eight feet and may weigh nearly 1000 pounds. The bonefish is a tropical species which is restricted to the warmer portions of Florida Bay near the tropical waters of the Keys.

Three members of the drum family are important temperate species in Everglades National Park. The black drum is the largest, but the red drum and the spotted seatrout (unrelated to freshwater trouts) are more numerous and among the most popular of all species with fishermen. All three occur around the entire Gulf coast and commonly up the Atlantic coast to Chesapeake Bay, with smaller

TABLE 13.1
Fishes of Importance in Florida Bay and Adjacent Marine and Estuarine Waters

Common name[a]	Scientific name[a]
game fishes[b]	
ladyfish	*Elops saurus*
tarpon	*Megalops atlanticus*
bonefish	*Albula vulpes*
common snook	*Centropomus undecimalis*
jewfish	*Epinephelus itajara*
crevalle jack	*Caranx hippos*
gray (mangrove) snapper	*Lutjanus griseus*
sheepshead	*Archosargus probatocephalus*
spotted seatrout	*Cynoscion nebulosus*
black drum	*Pogonias cromis*
red drum (redfish)	*Sciaenops ocellatus*
great barracuda	*Sphyraena barracuda*
Spanish mackerel	*Scomberomorus maculatus*
abundant nongame species[c]	
scaled sardine	*Harengula jaguana*
Atlantic thread herring	*Opisthonema oglinum*
hardhead catfish	*Arius felis*
inshore lizardfish	*Synodus foetens*
gulf toadfish	*Opsanus beta*
hardhead halfbeak	*Chriodorus atherinoides*
silverstripe halfbeak	*Hyporhamphus unifasciatus*
redfin needlefish	*Strongylura notata*
goldspotted killifish	*Floridichthys carpio*
rainwater killifish	*Lucania parva*
fringed pipefish	*Anarchopterus criniger*
dwarf seahorse	*Hippocampus zosterae*
gulf pipefish	*Syngnathus scovelli*
silver jenny	*Eucinostomus gula*
tidewater mojarra	*Eucinostomus harengulus*
pinfish	*Lagodon rhomboides*
striped mullet	*Mugil cephalus*
white mullet	*Mugil curema*
fantail mullet	*Mugil gyrans*
code goby	*Gobiosoma robustum*
clown goby	*Microgobius gulosus*

[a] Common and scientific names, arranged in phylogenetic order, follow Robins et al.,[203] except that vernaculars commonly used in Florida are included (in parentheses) for some game fishes.

[b] Selected from information by Tilmant.[233]

[c] Selected from studies conducted in Florida Bay.[218]

numbers found farther north. The red drum (locally called redfish and known as channel bass along the Atlantic coast) grows considerably larger, to 90 pounds and nearly five feet long in more northern parts of its range. The redfish population in the shallow waters of the park is made up mostly of individuals two years of age and younger and seldom as long as 30 inches. Like many other species, redfish use the marine and estuarine habitats of the area as a nursery, with larger individuals moving into deeper waters of the Gulf of Mexico.[234] Numerous studies of marine fishes have shown the importance of the nursery function of the area. The spotted seatrout spends its entire life cycle in the park's waters.[207]

Several migratory species occur in Everglades National Park. The Spanish mackerel is a temperate species that visits the park's waters in significant numbers in winter; it migrates from the northern Gulf of Mexico and from north of Cape Canaveral, where it resides during the summer.[108] It follows a pattern characteristic of a few other temperate species, which is the opposite of that exhibited by the tropical tripletail (*Lobotes surinamensis*). The tripletail drifts on warm currents near the surface and visits the park only in the summer. The tarpon also migrates, and large numbers of this spectacular game fish—often in excess of five feet long—move north for summer. Prime fishing for tarpon in the park occurs in April, at the beginning of the migration.

Mullet[210]

Mullet are abundant in Florida Bay as well as the other marine, estuarine, and even some fresh waters in Everglades National Park. Its importance is indicated by the fact that it is prey for every game fish listed in Table 13.1, except bonefish. It is often the first fish handled on a fishing trip because it is a popular bait.

Actually, four species of mullet occur in Everglades National Park, but the striped mullet is the most abundant. Mullets are essentially vegetarians, feeding by scraping algae and microorganisms from surfaces, by grubbing on the bottom, and by consuming surface film. They are an essential link in the food chain for game species and occur in enormous numbers in the park, where they also commonly enter fresh water. The striped mullet occurs almost worldwide in warm temperate and tropical inshore waters, especially in and near estuaries.

Importance of the Marine and Estuarine Fishes

To the sportsman, the marine and estuarine waters of Everglades National Park support an interesting mix of tropical and temperate fishes, composed of prominent migratory populations as well as permanent residents. The mix is attractive, demonstrated by the large number of visitors to the park whose only intention is a good day's fishing. However, the important ecological functions of the area are its support of fish-eating birds and its habitat as a nursery for a great many marine fishes, all based on the high productivity that arises from freshwater inputs from the Everglades as well as the neighboring Big Cypress Swamp ecosystems.[35] Wading birds utilize the mud flats of Florida Bay and the shallows of the mangrove swamps and other coastal habitats, and ospreys fish in the more open waters. All of these species prey heavily on the young of larger game fishes and on the high diversity of nongame species (the more common ones are listed in Table 13.1). Wading birds normally feed during low tide, which renders prey trapped and accessible and affords a daily food source for the species that typically live in marine and estuarine habitats. The marine and estuarine areas also are a critical alternative in years when little is available in fresh water for those birds that typically utilize the Everglades.

14

Amphibians*

T he biogeographic dispersal patterns of amphibians and primary freshwater fishes are very similar. The fact that all amphibians depend on fresh
water for at least the first stages of their life cycles accounts for this
similarity. Amphibian eggs are laid in fresh water, and larval stages, such as the
frog's tadpoles, live only in fresh water. Some amphibians, such as the large eel-
like sirens, are completely aquatic for their entire life histories. Others, such as
toads and treefrogs, live on land or on vegetation as adults and are able to move
from one stream system to another more easily than freshwater fishes. Nevertheless, amphibians have not dispersed to Florida via the overland route, northward
through Central America and around the Gulf of Mexico, or across salt water by
rafting on floating objects through the West Indies. As a result, no tropical
American amphibians have become naturally established in Florida.

The more common of the 15 native amphibians in Everglades National Park
are listed in Table 14.1. Most of these species occur throughout the southeastern
states, although many are secretive and seldom seen. Only the pig frog is restricted to the extreme southeast, although its range includes all of Florida. The
few types of salamanders that have been able to reach southern Florida are
naturally adapted to marsh and swamp habitats, such as the three-foot-long siren
and a distinctive Florida variety of the newt (*Notophthalmus viridescens*) which is
commonly found throughout the eastern states. Because of adverse environmental circumstances, none of the some four dozen woodland and brook salamanders
of the eastern states has been able to reach southern Florida. This restriction also
applies to most of the frogs and toads, with only a limited number of species able
to find habitats conducive to their survival. Thus, southern Florida has few native
amphibians, although many tropical species could live in the region if they could
get there. With human assistance, several exotic species have become successful
residents.

The squirrel treefrog may be the most commonly encountered amphibian for

---

* The primary citations for this chapter are References 8 and 133.

TABLE 14.1
Common Amphibians of the Everglades Region

Common name[a]	Scientific name[a]
greater siren	*Siren lacertina*
Florida cricket frog	*Acris gryllus dorsalis*
squirrel treefrog	*Hyla squirella*
green treefrog	*Hyla cinerea*
little grass frog	*Limnaoedus ocularis*
Florida chorus frog	*Pseudacris nigrita verrucosa*
eastern narrowmouth toad	*Gastrophryne c. carolinensis*
pig frog	*Rana grylio*
southern leopard frog	*Rana sphenocephala*
southern toad	*Bufo terrestris*
oak toad	*Bufo quercicus*

[a] Common and scientific names follow Ashton and Ashton.[8]

visitors to the park. Like the somewhat larger green treefrog, it can change color from green to brown, with several intermediate variations, including spots and blotches. During daytime, individuals are apt to crowd together in protected locations, often using man-made shelters. Their occasional loud, scratchy outbursts may mystify listeners. The location of the sound is elusive and, to the unfamiliar, indiscernible.

The squirrel treefrog and green treefrog, both highly photogenic as amphibians go, were once common in the Miami area, but have almost completely disappeared. Their elimination was probably due to predation by a much larger, introduced treefrog that is now well established in urban areas: the Cuban treefrog (*Osteopilus septentrionalis*). Oddly enough, a similar situation apparently resulted in the elimination of the southern toad from developed urban areas; the much larger marine toad or giant toad (*Bufo marinus*), which was introduced into the Miami area from Mexico, consumed the southern toad. The giant toad is highly successful in residential neighborhoods, where some unfortunate dogs meet their deaths every year after biting this species. Large poison glands located on the sides of the neck area exude a white, milky toxin when the toad is disturbed.

Because of its edible qualities, the pig frog is of economic interest (Figure 14.1). Its name derives from the close resemblance of its call to the casual, contented grunt made by farmyard pigs. Pig frogs are common, but are not easily seen because of their cryptic coloration and their tendency to stay away from shore; they remain dispersed throughout the deeper freshwater marshes. The pig frog grows very large, almost to the size of its close relative, the bullfrog (*Rana catesbeiana*), which is found farther north but does not reach southern Florida. Frog legs served in southern Florida restaurants are usually those of pig frogs,

FIGURE 14.1
A pig frog, the largest of South Florida's frogs, sits on a spatterdock leaf at the Anhinga Trail, Everglades National Park. (Photo by R. Hamer.)

but are known in the trade as bullfrogs for obvious reasons (legs from a "pig frog" might not market well, even though they probably taste and look the same as those of the widely appreciated bullfrog).

The Importance of Amphibians

The importance of amphibians in the Everglades ecosystem is inadequately documented and in need of research. However, while the list of species is short, amphibian populations are often so large that they must play a major role in the food chains. Evidenced by the din of their calls, most species start breeding upon the return of summer rains, and tadpoles immediately populate the newly flooded wetlands. Feeding principally on algae, they rapidly form a biomass of potential prey for fishes and wading birds. They are probably important foods for wading birds in short-hydroperiod wetlands and in isolated wetlands. In these areas, neither time nor access is sufficient for fish populations to invade and expand into a major biomass. In contiguous marsh areas of long hydroperiod, fishes form a much greater biomass, but even there, the tadpoles of some amphibians such as pig frogs, leopard frogs, and green treefrogs are common. In areas of very short hydroperiod (down to a minimum of about one month needed for the most rapidly developing species to mature from egg to young adult), larval amphibians may be more important than fish. Because many adult amphibians disperse from wetland habitats into uplands, they are also important to the food chains of uplands as well as wetlands.

15

Reptiles*

In contrast to amphibians, many reptiles can tolerate salt water and dry conditions. Sea turtles, for example, spend their entire lives in marine habitats, while alligators and some freshwater turtles can live in estuarine or even marine waters, at least temporarily. At the other extreme, many reptiles are completely terrestrial, spending their entire lives out of the water and even inhabiting deserts. Their ability to get along without a regular supply of drinking water far exceeds that of amphibians. A major factor in this independence from water is the reptilian egg; its hard shell and internal organization allow for incubation away from water. The live birth capability of a few (such as rattlesnakes) is similarly advantageous.

Reptilian tolerances for salt water and drought confer a superior potential over amphibians for crossing geographic barriers, but terrestrial and freshwater reptiles are still constrained by relatively small expanses of ocean. Only a few non-marine reptiles (namely some lizards) have invaded southern Florida from the tropics, probably by "rafting" on floating objects. Nearly all native, non-marine reptiles of the region came from the temperate land mass of North America, as did freshwater fishes and amphibians. Like those groups, several exotic reptiles have also been introduced by man. Miami and other urban areas now harbor about a dozen exotic species of lizards from tropical parts of the world.

Common or well-known reptiles of the Everglades region are listed in Table 15.1. All are allied to the temperate, North American continent except the American crocodile, the loggerhead turtle, and the green anole. Because of the secretive habits of many reptiles, relatively few are commonly seen by visitors to the Everglades. The American alligator—a grand exception—is easily seen because of its lifestyle and large size, and certain large turtles, notably the Florida redbelly turtle (Figure 15.1) and the Florida softshell (Figure 15.2), are also commonly seen basking or swimming in the clear waters of the Everglades (where their primary

---

* The primary citations for this chapter are References 7, 9, and 123.

TABLE 15.1
Common or Well-Known Reptiles of the Everglades Region

Common name[a]	General habitat(s)[b]	Scientific name[a]
crocodilians		
American alligator	f, b	*Alligator mississippiensis*
American crocodile	m, b	*Crocodylus acutus*
turtles		
loggerhead	m	*Caretta caretta*
common musk turtle (stinkpot)	f	*Sternotherus odoratus*
Florida mud turtle	f	*Kinosternon subrubrum steindachneri*
striped mud turtle	f	*Kinosternon baurii*
diamondback terrapin	b	*Malaclemys terrapin*
Florida box turtle	t	*Terrapene carolina bauri*
Florida chicken turtle	f	*Deirochelys reticularia chrysea*
peninsula cooter	f	*Pseudemys floridana peninsularis*
Florida redbelly turtle	f	*Pseudemys nelsoni*
gopher tortoise	t	*Gopherus polyphemus*
Florida softshell	f	*Apalone ferox*
lizards		
green anole (chameleon)	t	*Anolis carolinensis*
brown anole	t-i	*Anolis sagrei sagrei*
southeastern five-lined skink	t	*Eumeces inexpectatus*
snakes		
Florida green water snake	f	*Nerodia cyclopion floridana*
Florida (banded) water snake	f	*Nerodia fasciata pictiventris*
brown water snake	f	*Nerodia taxispilota*
peninsula ribbon snake	t, f	*Thamnophis sauritus sackeni*
southern ringneck snake	t	*Diadophis punctatus punctatus*
Everglades racer	t, f	*Coluber constrictor paludicola*
eastern indigo snake	t	*Drymarchon corais couperi*
Everglades rat snake	t	*Elaphe obsoleta rossalleni*
corn snake (red rat snake)	t	*Elaphe guttata guttata*
common kingsnake	t	*Lampropeltis getulus*
scarlet kingsnake	t	*Lampropeltis triangulum elapsoides*
Florida scarlet snake	t	*Cemophora coccinea coccinea*
eastern coral snake	t	*Micrurus fulvius fulvius*
Florida cottonmouth (water moccasin)	f	*Agkistrodon piscivorus conanti*
dusky pygmy rattlesnake	t	*Sistrurus miliarius barbouri*
eastern diamondback rattlesnake	t	*Crotalus adamanteus*

[a] Common and scientific names follow Ashton and Ashton.[7, 9]

[b] Type of habitat(s) of normal occurrence: b = brackish water, f = fresh water, m = marine, t = terrestrial; i = introduced.

FIGURE 15.1
A large Florida redbelly turtle in the jaws of an eight-foot alligator. The turtle, primarily herbivorous as an adult, is virtually helpless against the alligator once caught, but when the gator tried to break the shell, the turtle slipped away like a bar of soap and escaped—this time! (Photo by R. Hamer.)

FIGURE 15.2
A Florida softshell turtle, a predator primarily of invertebrates such as snails and crayfish, is an incredibly fast swimmer for a turtle, which is its primary defense against the alligator. (Photo by R. Hamer.)

concern is avoiding large alligators!). On the other hand, the highly venomous eastern coral snake and its diminutive and harmless near-mimics, the scarlet kingsnake and the Florida scarlet snake, are seldom seen. None of these snakes is common, and all tend to be nocturnal, living underground by day (fossorial). Other reptiles, such as rattlesnakes, are difficult to see because of their cryptic color patterns. Rattlers have become much less common, but are still a consideration when hiking overland. Even the pygmy rattler, whose adults are less than two feet long, can inflict a very dangerous bite. The list contains three commonly seen water snakes (Figure 15.3), which are nonpoisonous but resemble the less common and poisonous Florida cottonmouth.

Of the few tropical reptiles that have reached southern Florida by natural means, most are marine, including the American crocodile (discussed in a following section) and five species of sea turtles. The green turtle (*Chelonia mydas*) was formerly abundant, but is now an endangered species. The loggerhead is most often seen today and regularly nests on one of the park's few beaches, notably on Cape Sable. Of the terrestrial reptiles, only two are natural tropical

FIGURE 15.3
A Florida green water snake, one of several harmless water snakes that resemble the venomous Florida cottonmouth, at the Shark Valley Road, Everglades National Park. (Photo by R. Hamer.)

invaders of the region.* One is the reef gecko (*Sphaerodactylus notatus*), a very small and secretive lizard. It occurs in debris along beaches as well as inland in pinelands, hammocks, and in residential neighborhoods, but is seldom seen due to its nocturnal lifestyle. The other is the green anole, a common lizard that is also known as the *chameleon* because its color changes between green and brown. Its widespread occurrence throughout the South indicates that it has been there for some time and probably persisted in the area through the last glacial cycle.

Like many freshwater fishes and amphibians, southern Florida populations of several species of reptiles are distinct and recognizable as a variety or subspecies of their relatives living farther north. The subspecies of the common kingsnake has a very pale, speckled color pattern in contrast to those that live farther north in Florida or elsewhere in the range of the species. The Everglades subspecies of the rat snake is orange, in contrast to the dull yellow of most that live outside of the Everglades region. In addition, a number of reptiles, such as the Florida redbelly turtle, are confined to Florida even though they are closely related to species found farther north. These examples indicate that the environmental

* A third possible natural invader, disputed by herpetologists, is the bark anole (*Anolis distichus*), a small lizard that occurs in the Miami area, where it typically lives on tree trunks and cleverly keeps out of view on the opposite side from the observer.[217]

factors found in the Florida peninsula differ from those in the southeastern states and are favorable for special Florida or southern Florida varieties, or even separate (but closely related) species.

The gopher tortoise is unusual in the Everglades region, but is fairly common in other areas of Florida where sandy soils are dry enough for it to construct its deep burrow. A population of these tortoises is confined to the higher ground on Cape Sable[159] in Everglades National Park. This area is similar to an island in that it is isolated from the nearest other acceptable habitat by many miles of mangrove swamps and shallow tidal waters. The existence of the gopher tortoise on Cape Sable may be the result of human introduction, but it may also represent a relict population from times when sea level was lower and the habitat was contiguous with other areas.

The American Alligator[38, 69, 152, 238]

Most people would define an alligator as a large dangerous reptile found in the swamps of Florida. While that definition has some merit, it also contains some inaccuracies and ignores the fact that the alligator is the most important species in the Everglades. Those who love the sight of wading birds or enjoy fishing for bass in the Everglades owe a debt of gratitude to the alligator. In reality, driving a car to the Everglades is far more dangerous than encountering an alligator there.

Alligators and Other Crocodilians Compared

Alligators are part of a group within the reptilian class of vertebrate animals called *crocodilians.* There are at least 22 species* of crocodilians worldwide; 14 are true crocodiles, only 2 of which are responsible for the bad reputation attributed to all crocodilians. The culprits are the enormous Indopacific (or saltwater) crocodile of India, southeast Asia, and Australia and the Nile crocodile of Africa. Both of these species grow to be quite large, often exceeding 15 feet long.** Their

---

* The number of species recognized here is conservative; older literature often recognizes more. Differences lie in interpreting what constitutes a species, which is problematic for taxonomists working with most groups of plants and animals. Most often, it is geographic variation that may be interpreted in two ways: separate species or merely variation within a species. For example, the spectacled caiman, a South American crocodilian, has been counted here as one species (*Caiman crocodylus*), although it varies widely over its range and has been considered as several species by some authorities.

** Size records for crocodilians vary widely, in part because reptiles do not reach a terminal size as do mammals, but also because of exaggeration, which makes it difficult to verify the authenticity of records. However, to be conservative, the Nile crocodile reaches a maximum length of about 20 feet and the Indopacific can exceed 25 feet.[181, 205]

sheer size and technique of ambushing large terrestrial animals along the water's edge pose a considerable threat to man. Other types of crocodilians include the slender-snouted gharial (or gavial) of India (one species), the alligator-like caimans of South and Central America (five species), and the alligators. The American alligator of the southeastern United States and the smaller Chinese alligator are the only two types of alligators in the world.[181, 205]

In addition to the American alligator, the similarly sized American crocodile (see the next section) and occasionally an escaped caiman inhabit the Everglades. Most caimans are substantially smaller than the American alligator and crocodile. The common or spectacled caiman (*Caiman crocodylus*) is the species usually sold in pet shops and occasionally released by an unhappy owner who bought it as a novelty when it was only a foot long. It is very unusual to see a caiman or the secretive and rare American crocodile.

Size and Danger to Man

The largest American alligators ever known were about 17 feet long and the record is 19 feet 2 inches. If 12-foot alligators were common, the species might deserve its dangerous reputation, but in fact any alligator over about 11 feet is unusual, and just over 13 feet is the largest specimen found today.

In general, an adult human is too large to be of interest to an alligator, but precautions should be taken nonetheless. It is obviously dangerous to swim where a large alligator may be present, and children and even large dogs should be kept a safe distance from *any* alligator. Individual alligators accustomed to being fed are a particular danger. They become aggressive and associate people with food. For this reason, feeding alligators is illegal in Florida. Imagine an innocent visitor to a park, with toy poodle on leash, walking up to the pond for a closer look at the "tame" alligator. Through the gator's eyes, it looks like dinner is being served!

Alligators as Predators and as Prey

Raccoons are important predators on eggs and young gators. Otters, large-mouth bass, wood storks, large herons, and bull alligators add to the peril. However, gators longer than six feet appear to be safe from predation except by man and larger bull gators. In turn, gators attempt to catch and eat virtually anything that moves unless it appears too large. Juveniles feed mostly on invertebrates. In fresh water, snails and crayfish are important items on the menu, and in brackish areas, blue crabs are commonly taken. Gators over six feet long begin to take birds and mammals, including the species that would have preyed on them when they were smaller. Turtles and snakes—even poisonous species—are taken without hesitation and apparently without harm to the gator. Larger turtles normally escape until the alligators are large enough to have jaw strength capable of crushing the shell. Such strength is easily within the capability of a ten-foot gator.

Distribution

The American alligator (known in the vernacular as "gator") is found through-out all of Florida and Louisiana, as well as southeastern Texas, southern Arkansas, and coastal areas of Mississippi, Alabama, Georgia, and North and South Carolina. Although primarily an inhabitant of fresh water, it can tolerate brackish water and live for months in sea water. Its numbers have been reduced by land development, especially the draining of wetlands, and in the past by hunting for gator hides and even meat.

Protection of Alligators

Most people are familiar with the use of alligator hides for wallets and shoes, but few are aware that the meat, notably the tail, is considered a delicacy in some circles. In the 1950s and 1960s the rate at which alligators were disappearing was alarming. The market for hides supported innumerable gator "poachers." The rate of decline pointed to near certain extinction, at least for wild populations. Several states began prohibiting the taking of alligators, and the species was listed as endangered in 1966 under the federal Endangered Species Preservation Act. Initial provisions for enforcement against poaching proved ineffective however, and gators continued to disappear almost as rapidly as ever: policing the wilderness was fruitless. Subsequently, numerous state laws, a 1969 amendment to the federal Lacey Act, and finally the federal Endangered Species Act (1973) made it illegal to deal in any identifiable alligator "parts" (skulls, teeth, claws, meat, etc., as well as hides) in interstate commerce, thus providing the needed authority to enforce alligator protection. Products could no longer be legally sold, and the value of alligator hide plummeted. The population rebound of the alligator was rapid, first in Louisiana, then in Florida, and later throughout its range. As a result, the alligator was removed from the listing as endangered (except for purposes of its similarity of appearance to the American crocodile). However, alligator hunting is now strictly regulated through licensing, designated hunting periods and areas, and serial number tracking of hides.

The Alligator's Life Cycle

A walk through an alligator's life cycle will provide a better understanding of the importance of this complicated reptile in the Everglades. Males, or bulls, are much larger than females and are highly territorial during the mating season. A successful bull thwarts competition, sometimes in blood-chilling battles but usually by intimidation. Less aggressive bulls flee, an interaction that is commonly seen during the spring. Also, smaller alligators of both sexes are part of the diet of large bulls. While cannibalism may seem counterproductive to alligator success, it is really the opposite. It did not prevent the rapid rebound of alligator numbers when poaching was stopped, because large bulls were so rare that juveniles were almost free of a major predator. However, as the new bulls matured and became cannibalistic, they helped prevent overpopulation. This type of population control is an example of self-adjusting feedback.

Large bulls typically patrol territories of about five square miles. They wander over regularly used pathways, and studies have shown that they know where they are going, even as they travel long distances over land during dry periods. Authorities are often called upon to relocate potentially dangerous bulls, but some have been known to find their way back to their former territory; in one case a one-eyed individual returned 35 miles! Females, on the contrary, prefer home territory and are not prone to wandering great distances.

Female gators become receptive to males in April and May, near the beginning of the summer rainy season. Bulls can be heard "bellowing" at that time, although bellowing is not strictly confined to the mating season. The sound somewhat resembles the low rumble of the engine of a bulldozer. Bulls answer each other's bellowing, as well as loud noises from trucks and aircraft, especially sonic booms. The female occasionally bellows and will answer a bull's bellow with her higher pitched, softer voice.[25]

Alligators mate in the water, following a period of amorous nudging and cavortive swimming. After mating, bulls merely go on to the next gator hole in a polygamous lifestyle. A fertilized female begins building a nest by heaping soil and vegetation. Nests are usually built in marshes or on low tree islands near (but not next to) the gator pond (see next section). A nest may be as high as two feet above the water level. She lays an average of 30 eggs after the summer rains have reflooded the marshes in late June or early July. During the two-month incubation period, she normally tends the nest at night and leaves it alone during daylight. Young, "chirping," nine-inch-long gators are normally first seen and heard in late August in the Everglades. They are self-sufficient upon hatching, although the mother normally helps them out of the nest by gently tearing it open. An occasional female gator vigorously defends her nest against approaches by man, and there have been instances where intruders have been chased.

Female gators are tolerant of juveniles, even those of other families. It is not uncommon to see babies or even year-old gators sunning on the back of a female (Figure 15.4), but *never* on a bull. Gators grow at a maximum rate of about a foot a year, but growth depends on food supply and stress, and larger gators grow more slowly.

Alligator Holes and Their Importance

A female gator is a pond-builder. Given an area of wetland, she instinctively digs up vegetation and soil, piling the material around the edge of her excavation, and creates a pond called an *alligator hole*. It can often be recognized by the low trees and shrubs (often willows) that grow on the material pushed up around the edge of the pond. Gator holes also have two or three trails connecting them to adjacent marshes. Most gator holes are probably centuries old and have been maintained by successive generations of alligators. The pool of open water usually measures from six or seven feet to 20 feet across and is two to four feet deeper than the surrounding wetland. Gators also construct dens in the bank of the pond, with the entrance beneath the wet season water level.

It is a well-established fact that alligator holes are important to the success of

FIGURE 15.4
A female alligator with juveniles basking on her back at the Shark Valley Road, Everglades National Park. (Photo by R. Hamer.)

the Everglades as a haven for water birds. The relationship between alligator holes and wading birds is tied to the seasonal cycle of rainfall in the region. Late in the dry season, from November into May in normal years, when vast areas of the Everglades become dry, gator holes act as refuges for aquatic life.[137] Fish, frogs, turtles, water snakes, and other wildlife inhabit the ponds and use the alligator trails for access. Many become food for the resident gator, but they are also food for herons, egrets, ibis, storks, and anhingas. Furthermore, only in very severe droughts do the gator ponds become completely dry. Thus, some aquatic life normally remains to "seed" new cycles of life when water levels again flood the marshes and swamps during the wet season.* During extremely dry or cold weather, gators seem to disappear; actually, they retreat into their dens to aestivate

---

* Also, at the southern end of the Everglades where freshwater flows meet tidal water, numerous creeks provide a habitat that is very much like the alligator holes and, in parallel, is maintained by alligators. These creeks are the headwaters of the tidal rivers, such as the Shark River. Interconnecting alligator trails between these creeks and the adjacent marshes are thought to be similarly important to many species of wildlife (see Chapter 7 on mangrove swamps).[39]

(= estivate, or become dormant, a state exhibited by many animals in regions where adverse heat or drought recurs seasonally), leaving the gator hole to the other wildlife.

The preceding discussion has provided a broad description of the alligator in order to enhance understanding and appreciation of its importance. Consideration of this most interesting member of the Everglades wildlife should be at the heart of any discussion of the ecology of the Everglades.

The American Crocodile[159]

Unlike the alligator, the American crocodile (*Crocodylus acutus*) (Figure 15.5) in Florida is almost completely confined to the brackish and salt waters at the southern tip of the state, including the Keys. Lake Worth on Florida's lower east coast and Charlotte Harbor on the west coast are historically its northernmost limits. The species is also found in coastal habitats of the larger Caribbean islands as well as parts of Central and northern South America. Unlike the alligator, it is not adapted to cold weather and, therefore, could not survive in northern Florida, much less in other states of the southeast. Thus, the American crocodile can be categorized as tropical.

Adult American crocodiles attain roughly the same size as American alligators, but in general crocodiles are shyer; this is probably because they live in habitats where they do not often encounter people and therefore have less

FIGURE 15.5
A young American crocodile at Coot Bay Pond, Everglades National Park. (Photo by R. Hamer.)

opportunity to become accustomed to the presence of man.[151] Due to their preference for saline tidal habitats and their relative scarcity, crocodiles are rarely seen. Their coloration is brown to olive green, in contrast to adult alligators which are mostly black, with yellow banding particularly prominent in juveniles. The crocodile's snout, seen from above, is much more tapered and comes to a narrower tip. Also, the large fourth tooth, counting back from the front end along either side of the lower jaw, is conspicuously exposed on the outside when the mouth is closed. All the teeth of the lower jaw of the alligator fit into pockets in the upper jaw, leaving only the teeth of the upper jaw visible when the mouth is closed. Juvenile American crocodiles feed on fish and invertebrates such as crabs, while adults feed mainly on fish, of which mullet are an important species. The narrower jaw appears to be an adaptation for catching fish. Their technique of using a rapid sideways snap at potential prey is similar to that used by the slender-snouted gars and needlefishes.

Like alligators, crocodiles also build nests, but their nests are made of sand, marl, and peat piled up on dry land near water. Typically, the female lays 35 to 40 eggs during late April or early May, after which she checks the nest on regular nightly visits. About three months later, she assists the ten-inch hatchlings by digging them out of the nest and carrying them in her jaws to the water's edge; she remains with them for days or even weeks. Crocodiles do not build ponds as alligators do, at least in Florida; instead, they construct dens in the banks of tidal creeks or canals and create trails through mangroves. In general, less is known about the American crocodile than the alligator, and it certainly has less of an ecological impact in Everglades National Park due, in part, simply to its much smaller numbers.

The rarity of the American crocodile in southern Florida has caused wildlife management authorities to place it on federal and state endangered species lists. However, peculiarities of crocodile behavior make the species difficult to manage. Crocodiles often cross roads in their wanderings through coastal waters, resulting in road-kills. The section of U.S. 1 between the mainland and Key Largo has been a particular problem and is one of the few places on earth where "crocodile crossing" signs are posted to warn motorists. Most crocodiles remain within several miles of a "home" location, but some individuals under wildlife management surveillance have moved more than 60 miles. One specimen originally found off the beach at Fort Lauderdale was relocated to suitable habitat near Florida Bay. His subsequent travels included a visit to Big Pine Key, some 70 miles from the point of release. Such wandering of occasional American crocodiles through marine waters may explain how this species managed to reach southern Florida from the tropics.

16

Mammals*

Despite the great adaptability exhibited by mammalian evolution, temperate North America is the origin of all of Florida's native terrestrial mammals; none of the native species of the region came from the tropics. However, fossils prove that mammals from the American tropics invaded Florida during past interglacial times, following the emergence of the land bridge between South and Central America about 2.5 million years ago. All, however, perished during glacial periods when climatic conditions were cool and, more importantly, dry, thus eliminating their required wet forest habitat.[256]

The only possible exception to the lack of native tropical terrestrial mammals in Florida today is the nine-banded armadillo. This species is common in northern, central, and southwestern Florida, where sandy soils occur. Records of their occurrence in Everglades National Park are sporadic. Until fairly recently, the armadillo was only found in Central America and in some areas of the southwestern United States. Its movement eastward, across the Mississippi River and into Florida, is thought to have had considerable human assistance. Thus, its occurrence in Florida is generally regarded as another case of man's introduction of exotic species and not of a native terrestrial mammal.

Native Terrestrial Mammals

The common or noteworthy terrestrial mammals of the Everglades region are listed in Table 16.1. All but the round-tailed muskrat (or Florida water rat) occur widely in North America, some as far north as Alaska. The round-tailed muskrat is restricted to Florida and extreme southern Georgia and is much smaller than the common muskrat found elsewhere. Raccoons and marsh rabbits are probably

---

* The primary citations for this chapter are References 105 and 122.

TABLE 16.1
Common or Noteworthy Mammals of the Everglades Region

Common name[a]	Scientific name[a]
terrestrial and terrestrial/aquatic species	
opossum	*Didelphis virginiana*
marsh rabbit	*Sylvilagus palustris*
gray squirrel	*Sciurus carolinensis*
hispid cotton rat	*Sigmodon hispidus*
round-tailed muskrat	*Neofiber alleni*
gray fox	*Urocyon cinereoargenteus*
raccoon	*Procyon lotor*
Everglades mink	*Mustela vison evergladensis*
striped skunk	*Mephitis mephitis*
river otter	*Lutra canadensis*
panther	*Felis concolor*
bobcat	*Lynx rufus*
white-tailed deer	*Odocoileus virginianus*
marine/estuarine species	
Atlantic bottlenose dolphin	*Tursiops truncatus*
West Indian manatee	*Trichechus manatus*

[a] Common and scientific names follow Humphrey[105] and Burt and Grossenheider.[23a]

the most commonly seen mammals in Everglades National Park. The relatively small, dark-colored marsh rabbit often forages along roadsides and, unlike most other rabbits, is a willing swimmer. Otters, which are more at home in water than on land, are reasonably common but are usually seen only in the dry season, when they use roadside canals, particularly along the Shark Valley Road (Figure 16.1). The very rare, semi-aquatic Everglades mink is an endangered subspecies of the mink that lives throughout most of North America.

Although squirrels are probably the most commonly seen mammals in eastern North America, they are infrequently seen in Everglades National Park. Two ubiquitous species, the gray squirrel and the fox squirrel (*Sciurus niger*), live in southern Florida. A few gray squirrels are seen in the easternmost part of the park at Royal Palm Hammock. The Big Cypress fox squirrel (*S. n. avicennia*), a handsomely colored subspecies of fox squirrel also known as the mangrove or Everglades fox squirrel, inhabits the westernmost area of the park. Listed as threatened, its range is the vicinity of the Big Cypress Swamp, where it primarily inhabits pinelands but also enters hammocks and mangrove swamps.

White-tailed deer (Figure 16.2) are relatively common in southern Florida and are now much more common than they were historically because drainage has provided more dry land. It is not unusual to see deer feeding in the flooded

FIGURE 16.1
A river otter at the Shark Valley Road, Everglades National Park. (Photo by T. Lodge.)

marsh in the Everglades during early morning hours. White-tailed deer of the region are somewhat smaller than those that occur farther north, a geographic trend that culminates in the endangered key deer of the lower Florida Keys. Key deer are about the size of a large dog; they seldom weigh more than 80 pounds and stand only 25 to 30 inches high at the shoulder.

FIGURE 16.2
In the early morning, a doe and fawn white-tailed deer walk through a patch of wet prairie edged by a wall of sawgrass at the Shark Valley Road, Everglades National Park. (Photo by T. Lodge.)

The Florida Panther

By far the rarest mammal listed in Table 16.1 is the panther. It is known as the puma, cougar, or mountain lion in western parts of North America where it also lives. The endangered Florida population, which is considered to be a separate subspecies (*Felis concolor coryi*), is currently composed of 30 to 50 individuals—all confined to southern Florida. At the time of this writing, however, none of these large, light-brown cats remains in Everglades National Park. Adults may grow to seven feet, including the tail, and weigh from 65 (small female) to 160 pounds (large male). Their footprints, like those of other cats in that they lack claw marks, are over three inches in diameter, which is far larger than footprints of the moderately common bobcat. The home range of a panther is typically about 250 square miles for males and less than 100 square miles for females. Travel is normally during the night, but in cooler winter weather they may also move during daylight. Panthers are mostly solitary, except for the 12 to 18 months when a female is raising her one to four kittens. An adult male or female is generally intolerant of the presence of another adult panther of the same sex in its territory, and intrusion occasionally results in a fight to the death.

The Florida panther's preferred prey is deer, but any small to medium-sized animal may be taken, including rodents, rabbits, raccoons, and even small alligators. This panther is not considered to be dangerous to man and is currently the focus of an intensive program to understand, protect, and reestablish populations in Florida. Many individuals have been fitted with a collar containing a radio transmitter. This program has received much criticism by adversaries who feel that the animals should not be disturbed. The objective of the program is to monitor panthers' travels in order to better understand and protect this nearly extinct subspecies. The information has helped, for example, in improving highway design in critical areas, most notably Interstate 75 through the Big Cypress Swamp, so that road-kills are less likely. Wildlife underpasses and roadway fencing have proved effective for panthers as well as other wildlife. A proposed captive breeding program is controversial and is considered to be a last-ditch effort to prevent extinction.

Marine Mammals

Two marine mammals of note in the nearshore waters of southern Florida including the shallow bays and estuaries of Everglades National Park are the Atlantic bottlenose dolphin and the West Indian manatee (an endangered species). Numerous other marine mammals would be listed if offshore waters were included, but most of these are rarely seen. It is also of interest that the West Indian monk seal (*Monachus tropicalis*) once occurred in marine waters around peninsular Florida and the Keys, but it was hunted to extinction during this century. Bottlenose dolphins are common in the waters of the park, where they feed on fish. The manatee, by contrast, is a rather sluggish vegetarian and is one of the few marine mammals that can be said to be tropical. It is sometimes killed by cold water farther north in Florida, and in some places it now takes advantage of heated water from power plants during the winter months. Manatees commonly enter fresh water, and some individuals occasionally reach Lake Okeechobee. The status of the manatee is now listed as endangered. The primary threat is boat traffic, and many individuals show the telltale series of scars from encounters with boat props.

17

Birds*,**

W ith close to 400 species of birds having been documented as occurring naturally in southern Florida (about 350 specifically within Everglades National Park), the overall list of birds, or *avifauna*, is rich. Within the total avifauna, just under 300 species are considered to be of regular occurrence, while other species are known from only a few records.

Before purchasing binoculars and a field guide to birds, some details of this rich avifauna are in order. About 60 percent of these species are winter residents, migrating into South Florida from the north, or are hurried visitors, stopping only briefly in the spring or fall in their migration between tropical America and points farther north in the United States or Canada. The remaining 40 percent of the species of birds that regularly occur in southern Florida are the true natives—the birds that breed there. Breeding range can be thought of as the true home of a bird species from the standpoint of biogeography.

The list of species of birds breeding in southern Florida, which includes about 116 species, is substantially smaller than lists for areas of similar size in the northeastern United States. Furthermore, not all of these South Florida breeders are year-round residents. For example, wood storks breed in southern Florida in the winter and spring and then move farther north in Florida and adjacent areas for the summer. In an alternate lifestyle, the exquisitely graceful American swallow-tailed kite breeds in the region in late spring, then leaves for tropical America in August and does not return until February. For most people, however, the dry

---

* The primary citation for this chapter is Reference 200.

** Common names of birds have been so meticulously standardized in periodic checklists by the American Ornithologists' Union (which most guidebooks follow, e.g., Reference 162b) that the use of scientific names is not necessary in this text. As a result of standardization, authorities have decided that the names are proper nouns and should be capitalized, but there is wide disagreement on this point. For the sake of uniformity in this text, they will not be capitalized (it is standard practice not to capitalize common names of plants and animals).

season is an excellent time to see birds in southern Florida. The optimum is from mid-December into April.

An analysis of the geographic origins of the breeding birds of southern Florida reveals some interesting trends. Given the ability of birds to fly long distances over water, one would think that the region, with its predominance of tropical vegetation and proximity to tropical areas such as Cuba, would also be a haven for tropical birds. It may come as a disappointment that this is not generally the case, but the tropics have indeed enriched bird life in southern Florida. To explain the situation, it is helpful to discuss South Florida's breeding birds by dividing them into two categories: land birds and water birds.

The distinction between land birds and water birds has to do with habitat requirements, rather than their relationships by evolution. Natural groupings of birds tend to include species that are mostly all either land birds or water birds, and therefore the land and water categories have few misfits. Land birds are those types of birds that normally live in terrestrial habitats, without particular dependence on aquatic or marine habitats. Examples of groups of birds that can be categorized as land birds are hawks, hawk-like birds including vultures and kites, doves, owls, woodpeckers, and songbirds (sometimes called perching birds, but technically known as the passerine birds). The songbirds or passerine birds include many species such as swallows, jays, the mockingbird, thrushes, flycatchers, warblers, and black birds. It is convenient to distinguish between passerine and non-passerine land birds in some comparisons.

There are some misfits within this land bird category. Obvious examples are the osprey (or fish hawk) and the bald eagle, both of which are seen regularly in the Everglades. Both are associated with water but are included as land birds. However, it is difficult to find more than a handful of exceptions, and they are unimportant for the purpose of this discussion.

Water birds, then, are those types of birds which are strongly associated with freshwater, estuarine, or marine habitats. They include grebes, pelicans, cormorants, the anhinga, ducks, herons and egrets, ibises, the wood stork, cranes and their relatives such as gallinules, rails and the limpkin, shore birds, gulls, and terns. Again, there are only a few exceptions. The cattle egret, an insect eater, prefers dry grasslands and an association with cattle or even noisy tractors to help stir up a meal. The killdeer (in the shore bird group by natural relationships) prefers the open fields, rather than wetlands or shores like its relatives. Nevertheless, these few exceptions are considered together with the water birds.

Breeding Land Birds

Just over 70 species of land birds breed in southern Florida, half of which are songbirds or passerine types. Some interesting geographic trends in these land birds are shown in Table 17.1. It is obvious that southern Florida has relatively few breeding land birds, and the increase in numbers when comparing southern Florida to all of Florida and then to comparably sized states farther north (Ken-

TABLE 17.1
Geographic Trends in Numbers of Breeding Land Birds in the Eastern United States[a]

Region	Numbers of species		
	Breeding passerines (songbirds)	Breeding non-passerines	Total breeding land birds
South Florida	36	37	73
Florida	61	43	104
Kentucky	85	34	119
Michigan	109	43	152

[a] Modified from Robertson and Kushlan.[200]

tucky and Michigan) is in passerine birds. The number of species of remaining non-passerine land birds in each region is about equal.

Within the Florida peninsula, there is a pronounced decrease in the number of breeding passerine birds from north to south. Part of this trend is understandable because of habitat limitations. In southern Florida, wetlands (especially the Everglades) account for most of the area, which leaves little habitat for land birds. However, the breeding ranges of many land birds end in areas where habitat does not appear to be limiting. Furthermore, a careful evaluation of available data has shown that the breeding ranges of numerous species have been moving northward, farther and farther from southern Florida. It is thought that much of this northward movement is not due to human influences.

A hypothetical explanation is that this northward shift is a continuing biological response to the overall climatic changes that ended the last glaciation. These particular passerine birds of temperate origin were probably much more numerous in Florida during the height of the glacial age when Florida was much larger, cooler, and drier than today. Presumably, their populations have been shifting slowly northward over the past 15,000 years, in response to climatic change. Further, there is abundant evidence in Europe and North America that the northern breeding limits of many species of birds have been moving north over the years. However, the hypothesis is complicated by several species, including the indigo bunting, gray catbird, and American robin, whose breeding ranges have been moving southward in Florida in recent years.

With relatively few breeding land birds present in southern Florida, why have tropical species not moved in from Cuba, Hispaniola, and the Bahamas? Very few species of tropical land birds are naturally present in southern Florida even though substantial, but not rich, land bird faunas are present in these adjacent West Indian islands. Before the turn of this century, Florida had a large population of the Carolina parakeet, a beautiful, small, yellow and green parrot species. Its range actually extended throughout the eastern half of the United States. This species, now extinct due to hunting, represented the single success of the parrot group in naturally reaching Florida. On the other hand, a few tropical land birds are naturally expanding populations and breeding ranges in Florida. Notable examples are two non-passerines (the smooth-billed ani and the white-crowned pigeon) and two songbirds (the gray kingbird and black-whiskered vireo). Other species also appear to be ready to make the move, with occasional visits. The process has been slow and has resulted in an apparent deficit of breeding land birds in southern Florida.

One last word on this topic: people have successfully introduced a large number of tropical land birds into southern Florida, including the spot-breasted oriole, the red-whiskered bulbul, and several parrots (red-headed amazons, monk parakeets, and canary-winged parakeets lead the list). The success of these species is likely helped by the weak competition resulting from the relatively few native species of breeding land birds in the region and their adaptability to man's altered and disturbed habitats.

Breeding Water Birds

A selection of common or well-known water birds in the Everglades ecosystem is provided in Table 17.2. In contrast to land birds, more species of breeding water birds are listed for Florida than for states farther north (see Table 17.1). Southern Florida alone boasts almost 120 species, with 43 breeding in the region. Breeding water birds of southern Florida differ markedly from the land birds in both geographic distribution and general origin. First, the number of breeding species is the same for southern Florida as for the remainder of the state, although the list of species differs somewhat. A trend of decreasing numbers begins north of the state.

Second, a large percentage of the water birds that breed in southern Florida are also common to the West Indies. These species are generally considered to be tropical, although many have ranges that extend far north into the continent. Species that prefer shallow tidal habitats appear to treat southern Florida just like another tropical island. The greater flamingo of the Caribbean region occasionally visits the shallows of Florida Bay (but most of the few flamingos seen in the past few decades are thought to have escaped from captivity). The reddish egret, found in the shallow estuarine and marine waters of the region and especially Florida Bay, also uses shallow tidal waters around islands of the West Indies and Bahamas. The roseate spoonbill, which frequents similar habitats, moves about the region in an opposite pattern, with many individuals migrating northward from Cuba to southern Florida to breed during the winter months.

Pelagic or open-ocean sea birds also treat southern Florida as another Caribbean island. Examples that breed in the region include the magnificent frigatebird and two kinds of terns. However, several other pelagic sea birds regularly seen at sea off the coast could potentially nest in southern Florida except for lack of proper habitat. Florida has none of the rocky cliffs preferred by many pelagic sea birds. The use of such inhospitable nesting sites helps to prevent predation of

TABLE 17.2
Selected Common or Well-Known Water Birds of the Everglades Ecosystem[a]

Common name[b]	General habitat(s)[c]	Scientific name[b]
diving birds		
pied-billed grebe	f	*Podilymbus podiceps*
anhinga	f	*Anhinga anhinga*
double-crested cormorant	f, m*	*Phalacrocorax auritus*
common merganser	f*, m	*Mergus merganser*
red-breasted merganser	f, m*	*Mergus serrator*
wading birds		
least bittern	f	*Ixobrychus exilis*
American bittern	f*, m	*Botaurus lentiginosus*
black-crowned night-heron	f*, m	*Nycticorax nycticorax*
yellow-crowned night-heron	f, m*	*Nyctanassa violacea*
green-backed heron	f, m	*Butorides striatus*
tricolored heron	f, m	*Egretta tricolor*
little blue heron	f*, m	*Egretta caerulea*
reddish egret	m	*Egretta rufescens*
snowy egret	f, m	*Egretta thula*
great egret	f, m	*Casmerodius albus*
great blue heron	f, m	*Ardea herodias*
great white heron	f,m*	*Ardea herodias*
wood stork	f*, m	*Mycteria americana*
glossy ibis	f*, m	*Plegadis falcinellus*
white ibis	f, m	*Eudocimus albus*
roseate spoonbill	f,m*	*Ajaia ajaja*
limpkin	f	*Aramus guarauna*
birds of prey[d]		
bald eagle	f, m	*Haliaeetus leucocephalus*
snail kite	f	*Rostrhamus sociabilis*
osprey	f, m	*Pandion haliaetus*
other birds of interest		
purple gallinule	f	*Porphyrula martinica*
common moorhen	f	*Gallinula chloropus*
American coot	f	*Fulica americana*
Cape Sable seaside sparrow[d]	f	*Ammodramus maritimus mirabilis*

[a] Most species are selected for their dependence on fish or aquatic invertebrates of the Everglades. Exceptions, such as the Cape Sable seaside sparrow, are included because of special interests mentioned in the text.

[b] Common and scientific names follow National Geographic Society.[162b]

[c] The type of habitat(s) of normal occurrence: f = fresh water, m = marine and estuarine; asterisk (*) indicates the more common habitat.

[d] Birds of prey as a group and the Cape Sable seaside sparrow are not categorized as water birds, but these particular examples are associated with aquatic habitats.

eggs and nestlings by land animals, in particular the raccoon. Because the magnificent frigatebird is accustomed to using mangrove trees on isolated islands as nesting sites, it finds southern Florida acceptable, although actual nesting is restricted to the westernmost keys.

Of the species with ranges that extend far north into the continent, most swell the southern Florida population with wintering individuals. The great blue heron ranges as far north as central Canada during summer months. While some remain relatively far north in winter, most migrate to tropical America. A substantial number winter in southern Florida, together with resident Florida individuals and other similar migrants.

Two species of pelicans can be seen in southern Florida. The more common brown pelican breeds in Florida and is generally tropical. It seldom ventures inland and feeds on fish by diving into the water head first. Its lower beak supports a huge throat pouch which expands like a parachute on hitting the water. How the bird manages this technique without breaking its thin neck is mystifying and is in marked contrast to the technique of the American white pelican. The latter species does not breed in southern Florida but winters there, spending its summers mostly in freshwater habitats of the western United States and Canada. White pelicans feed while swimming and frequently cooperate in large groups. They are an exciting addition to the wintering wildlife of the region.

The superficially similar double-crested cormorant and anhinga add only two more breeding water birds to the southern Florida list, but they are particularly interesting fish-eating birds. The cormorant is primarily a coastal species and catches fish in its beak during underwater dives. It is easily distinguished from the anhinga by the prominent hook on its beak. The anhinga (sometimes called the snakebird, water turkey, or darter) normally uses interior fresh waters, where it is frequently seen with head and snake-like neck alone protruding from the water. This species catches its prey by spearing it with its long, pointed beak. When successful, it surfaces holding the impaled fish in the air. Then, with a quick motion, the anhinga flips the fish up and catches it in the beak. A few adjustments are made to maneuver the fish, and it is swallowed head first. The use of the beak as a dagger is a chilling reminder of how dangerous some wildlife can be if cornered or approached when injured. This bird, as well as the herons and egrets, can put out an eye with one sudden jab. Left alone, these birds are harmless and highly entertaining in their peculiar feeding styles.

The anhinga is probably the most memorable water bird seen in Everglades National Park, and the most popular boardwalk viewing area in the park, the Anhinga Trail, is its namesake. The anhinga's body feathers are structurally incapable of trapping air, unlike most diving birds. Consequently, a diving anhinga's buoyancy is close to neutral. These birds are able to swim about slowly under water—through vegetation and around obstacles—with little effort. However, the wet feathers are very poor insulators, and a great deal of body heat is lost during their underwater work as well as through evaporation afterwards. This leads to a commonly stated misconception. In a highly conspicu-

FIGURE 17.1
A female anhinga thermoregulates after a dive: warming up, not drying out. (Photo by R. Hamer.)

ous pose, anhingas characteristically perch with wings outstretched after each underwater excursion (Figure 17.1). People commonly remark that they are "drying their wings so they can fly." While this pose aids in drying, its importance is in raising the bird's body temperature or *thermoregulation*. Special blood vessels in the wings act like water pipes in a solar collector, capturing the sun's warmth. Furthermore, anhingas can fly even when soaking wet. Given the right stimulus, such as an approaching alligator, an anhinga can burst from the water into flight.

Ducks and shore birds are very poorly represented in southern Florida as breeding populations. By stretching the numbers to include infrequent occurrences such as the breeding records of the American oystercatcher, there are only about six shore bird species and three ducks, including the resident but secretive wood duck. However, wintering species in each of these groups number about 20.

Feeding Behavior of Wading Birds[93, 111]

The large wading birds of the Everglades region attract much attention. Fourteen species breed in the region, and their feeding activities are of particular interest, each with its own specialized style. A short diversion into some theoretical ecology will add insight in understanding why each species is unique.

Ecologists have rather abstractly defined the entire scope of activities of a given species as its ecological niche. This niche is not just a place (the term *habitat* is used for the place or places used by a species). Instead, an ecological niche is more comparable to a profession, although the habitat where the profession occurs is part of the picture. It is theorized that direct, total competition between two species (the condition in which they share an identical niche) would result in the disappearance of one species. This concept is called *competitive exclusion*, and it probably explains why some introduced species are unable to become established in a new region even though the available habitat appears appropriate for their survival. Severe competition from one or more established species is usually thought to be the negative factor in these situations. In ecological terms, the required niche is already filled by one or a combination of species established in the region.[167]

While competitive exclusion probably does occur, species can and do change through the evolutionary process of adaptation by natural selection. The world is full of examples where natural selection appears to have changed the physical shape, size, or behavior of species. A famous showcase of such adaptations involves the finches of the Galapagos Islands, located in the Pacific Ocean off the coast of Ecuador.[90] Inheritable variations that enable the individuals to use a slightly different niche reduce competition. Such advantageous variations improve the *fitness* of an individual, which is defined by its chances for survival and reproduction; thus, the changes become more common through succeeding generations (i.e., the species evolves). For example, a change may allow a species to obtain slightly different foods not readily available to competitors. This specialization would be an adaptive change in the niche of the species. On careful observation, it is usually easy to identify physical and behavioral specializations that reduce competition among even very similar species.[90, 148]

Few areas on earth allow a casual observer to see the subtle differences between competitors better than in the feeding behavior of wading birds in Everglades National Park. The park offers this unusual opportunity because of the large number of species that can be seen and because they can be observed closely (the park's protection makes many of the birds unafraid of people). The best opportunities for observing wading birds arise when drying conditions force large concentrations of food (mostly fish and aquatic invertebrates) into shallow ponds, and huge numbers of water birds descend on the concentrated prey. At these times, the birds are forced into their highest level of competition, but specializations are still easily seen. The Anhinga Trail and the Shark Valley Road area are known for their particularly long periods of wading bird activity. Numerous other locations offer very good opportunities during this period, with optimum times differing from place to place and year to year. A famous but short annual feast occurs at Mrazek Pond, not far from the park's facilities at Flamingo. The pond lies at the edge of the road, which makes it easily accessible. Activity generally lasts only a few weeks each year, with the heaviest use lasting only about a week. It may occur any time between late December and April, depending on water levels.

The most obvious difference in feeding techniques that reduce competition is the touch-oriented feeding of the wood stork, white ibis, and spoonbill versus the visually oriented feeding of the herons and egrets. Some behavioral traits of several prominent Everglades species will further explain these differences.

Wood Stork[111, 171, 244]

Of the three species that locate prey by touch, the wood stork commands respect as the true fisherman; it takes mostly fish and few invertebrates. Often feeding in cooperative groups, its technique, called groping, is to probe its long, heavy, partially opened bill into the water while walking slowly about (Figure 17.2). If a living organism is detected by the highly sensitive receptors on the inside of the bill, the bill snaps shut by reflex action in less than 0.03 seconds. If something is caught, it is picked up and, with a quick snap, tossed to the back of the mouth and swallowed. While most of the bird's motions are relatively slow, the catch-and-swallow sequence is often so fast that one must watch very closely to see what is being eaten. Only when the prey is large, or spiny such as a catfish,

FIGURE 17.2
A wood stork groping for a fish in Mrazek Pond, Everglades National Park. (Photo by R. Hamer.)

does the stork take some cautious preparations. Normal feeding behavior also involves assistance with one of its light pink feet, using a slow stirring motion that disturbs any motionless fish. In murky or weed-choked water between 6 and 20 inches deep, wood storks have a considerable advantage over sight-oriented wading birds. On the other hand, the technique is not effective in clear water where potential prey can see and evade these large, awkward birds. The stork's alternative sight-oriented feeding behavior is inefficient compared to herons and egrets.

Most of the wood stork's diet consists of fish which measure between two and ten inches long, but it also eats a variety of other creatures, including larger invertebrates and sometimes even baby alligators. Wood storks may also feed effectively even when all surface water is gone and prey is concealed in soft mud.

The specialized grope feeding behavior of the wood stork has rendered it the most endangered of Florida's wading birds. It is highly dependent upon a concentrated food source. Nesting failures result in years that lack dry weather or when water is kept too deep by artificial management. Further, nesting also fails in years when the drydown conditions proceed too rapidly, so that the required food source is largely depleted before the wood stork young are fledged. An orderly sequence of drydown in freshwater wetlands is the key to success, but this is no longer the norm in the modified Everglades (see Chapter 18). Breeding populations of the wood stork may disappear in the Everglades region unless restoration plans are successful.[172]

White Ibis

The narrow, downward-curving beak of the white ibis is an indicator that something is unique about its mode of feeding. In contrast to the wood stork's slow, cautious approach, the white ibis rapidly probes the water—sometimes with its head completely submerged—and explores in, around, and under obstacles. The result of these efforts nets a much higher percentage of invertebrates, typically crayfish and insect larvae in fresh water and small crabs in saline coastal areas. Much of this prey is extracted directly from burrows or other hiding places. Most of the relatively few small fish caught by this technique are probably those that have taken shelter in some nook or cranny where they are safe from other predators—but not from the white ibis.

Glossy Ibis

A puzzling question in comparing ecological niches is the similarity between the white ibis and the dark-colored glossy ibis, which is more common farther north in Florida and along the Atlantic coast. Its populations fluctuate in southern Florida, but it frequently visits the northern portion of Everglades National Park. While the slight difference in the ranges of these two species may constitute a difference in their respective niches, skillful observation could probably also reveal some subtle differences in feeding techniques. Both of these species also feed on land, including fields and golf courses, especially following heavy rains.

The glossy ibis is locally abundant, and its populations have been expanding since it first started breeding in Florida in the 1880s. The white ibis is the most abundant wading bird in Florida, but its population has recently declined alarmingly while the glossy has increased (competition?).[111, 206]

Roseate Spoonbill

The bizarre spoon-shaped beak of this species obviously indicates that its feeding technique differs radically from other wading birds (Figure 17.3). The spoonbill's normal feeding behavior is to swing the mostly submerged opened bill in a wide arc from side to side as the bird walks about in shallow water. When several birds are present, they often team up, forming a line. The mature spoonbill's brilliant pink wings, white back and neck, and pale greenish head make these cooperative feeding efforts a sight to behold. The technique is successful in catching small fish and invertebrates at mid-depth in the shallow water column, or even strained from loose sediments at the bottom. Various small shrimp, called prawns, are an abundant component of their diet. Spoonbills nest primarily on mangrove islands in Florida Bay, but normally fly to brackish wetlands of the mainland to feed. The juvenile birds are very pale (almost white) and are thought to obtain their adult pink color by extraction of a red pigment (related to vitamin A) from certain crustaceans in their diet.

Great Blue Heron

A great blue heron was once seen impaling a 16-inch largemouth bass, perhaps weighing three pounds, which illustrates the deadly force with which this sight-oriented heron can attack its prey. That event at the Anhinga Trail in Everglades National Park was unusual, and an opportunistic alligator ended up with the prize, which the heron was barely able to carry, much less eat. The technique of spearing fish from 3 to 12 inches long is common for "great blues," which are most abundant in southern Florida in winter (when migratory individuals join the resident population) in both interior freshwater wetlands and in saline coastal habitats. This species objects to the close presence of other members of its species and tends to feed alone. Aided by its large size, the great blue heron often fishes in water up to about two feet deep, which is much deeper than other herons except the closely related great white heron. If attracted by abundant prey, great blues will even swim—like ducks—in water too deep to touch bottom.

Great White Heron

There is an ongoing disagreement among ornithologists as to the status of this bird: whether it is merely a color form of the great blue or a separate species. Whatever the case, the great white heron is almost completely confined to saline habitats of southern Florida and the West Indies. Close to extinction in Florida after the 1935 Labor Day hurricane, it has rebounded well, but Hurricane Donna

FIGURE 17.3
A bizarre mouth in the Everglades, the roseate spoonbill at Mrazek Pond, Everglades
National Park. (Photo by R. Hamer.)

(1960) set it back somewhat. It is now commonly seen, usually standing alone in the shallows of Florida Bay. Its habitat specialization makes its niche quite different from that of the more ubiquitous great blue. These two large herons occasionally interbreed and produce offspring that were once thought to be yet another kind of heron, called Wurdemann's heron, recognized as being like a great blue but with an almost pure white head and neck. Wurdemann's herons are most often seen in the Florida Keys.

During the spring, juvenile and some adult great white herons may move northward from Florida Bay into the freshwater Everglades. At the Anhinga Trail, located in a freshwater area of Everglades National Park, a voracious individual was observed while spearing and swallowing about 20 six- to eight-inch golden shiners, which were abundant that year (see front cover). On other days, it caught and consumed a dragonfly and a small turtle and swallowed the remains of a dead and badly deteriorated anhinga, complete with beak intact. In dry coastal habitats, great whites have been seen catching and eating hispid cotton rats. Thus, while most great whites live and feed in Florida Bay and eat fish similar in size to those selected by the great blue, they will also eat almost anything else they can swallow. This ability to improvise in unusual situations is part of the niche of this species. Many species lack such versatility and such voracity.

Tricolored Heron

Formerly called the Louisiana heron, this species ranges widely over the Caribbean region, throughout Florida, and into closely neighboring areas of the southeast (Figure 17.4). Farther north it is most common in saline tidal areas, but in Florida it also frequents freshwater habitats. Compared to the great blue heron, this smaller, beautifully colored species is a restless and agile athlete, actively running after the inch-long fish that form the bulk of its diet. Its preferred habitat is a wide expanse of very shallow water; two to five inches deep is probably optimal for its highly active style. Tricolored herons normally feed alone but will take advantage of disturbances caused by other animals. Along a tidal creek near Mud Lake, a tricolored was seen walking quickly along the shore while a red-breasted merganser (a fish-eating duck) swam a parallel course just offshore. The small fish between the two were in a state of panic. Both birds benefited from the confusion. Sailfin mollies and killifishes are perhaps the most common prey of the tricolored heron. The exquisitely colored breeding adults, with blue beaks and dark, ruby-red eyes, are the most striking in appearance among southern Florida's herons and egrets.

Reddish Egret

Slightly larger than the tricolored heron, the reddish egret is a specialist of shallow marine flats at low tide and is seldom found inland. It comes in two color phases: a generally rusty-brown form and a white form, which is easily confused from a distance with the smaller snowy egret and immature little blue heron. The

FIGURE 17.4
In breeding plumage, a tricolored heron hunts through an open marsh of spike rush at Nine Mile Pond, Everglades National Park. (Photo by R. Hamer.)

antics of this highly active species include the athletic ability of the tricolored heron in addition to the regular use of wings as a sun visor for shading and for maneuvering. Its athletic performance, however, requires fairly regular rests, during which the bird stands motionless for a short time before resuming its agile endeavors. Experienced observers can recognize the reddish egret at a distance purely by its behavior. When this species "fluffs" its feathers, the display of shaggy head, neck, and back feathers is quite a surprising treat. During feeding, these feathers are normally held flat against the body. The species' decorative feathers brought it closer to extinction by plume hunters than any other species. Its recovery since the early part of this century has been very slow.

Great Egret

Through its series of official name changes from the misnomer *American egret* (the species is found not only throughout most of North, Central, and South America but also in Europe and much of Africa, Asia, and Australia!) to *common egret* (another misnomer because it was almost hunted to extinction in North America because of its beautiful plumes) to the current *great egret,* the bird has remained the same. *Great patient egret* would be a more descriptive name for this species. Periods of nearly motionless stance followed by a slow, careful walk describe its normal style. It prefers to feed alone in open marsh habitat. However, in times when food is concentrated into pools by low water, it will feed near others of its own species, as well as other kinds of water birds. Its posture of leaning far forward, with neck outstretched and head held horizontally rather than pointed downward like most herons, is typical. In final efforts of locating prey, it is apt to move its head from side to side, as if to enhance its depth perception. It eats mainly fish, but also takes amphibians, reptiles, and small mammals when the opportunity arises. Generally, its prey is substantially smaller than prey selected by the larger great blue and great white herons.

Snowy Egret

What the snowy lacks in size, this smaller heron offsets by aggressive feeding. Yellow slippers and fast footwork are the trademarks of this gregarious species, which is commonly seen feeding in flocks with numerous other species. Its behavior varies more than any of its competitors. When wading in shallow water, this swiftly reacting bird often use its feet in creative ways: stamping, stirring, and probing to flush out potential prey or even chattering its bill in the water. Some of these actions attract its principal food, the mosquitofish, an inch-long species that is apt to investigate disturbances in seeking its own food. At times, a snowy will search slowly and carefully, but more often it acts nervously, turning quickly and jabbing at prey, especially when competing with other water birds. It may even feed on the wing, similar to a sea gull.

Little Blue Heron

The most characteristic feeding behavior of the little blue can be described as meticulous investigation. This species is most at home along the edges of bodies of water, where it walks very cautiously and carefully examines any obstacles such as stones, plants, and logs. While the little blue will feed competitively with other herons in shallow, open water, it looks awkwardly out of place, like it has been pushed into an activity for which it has limited enthusiasm or training. The little blue's preferred techniques give it a strikingly different diet than the previously described species. Small frogs and polliwogs, insects, crayfish, prawns, spiders, and other invertebrates form a much larger proportion of the little blue's diet than is normal for the other herons and egrets. Much of its prey is caught on vegetation above water level, but like all the other wading birds, it will take available fish. Identification of juveniles is tricky because they are white and resemble snowy egrets.

Green-Backed Heron

With limited observation, the green-backed heron (formerly just the green heron) looks like a bird with no neck. Even experienced observers are surprised when this squat little heron suddenly thrusts forth a neck far longer than conceivable from the well-concealed S-shaped coil beneath a cover of feathers. Its legs are relatively short for herons, but are deceptively longer than they appear during the bird's normal crouched stance. Found in numerous varieties throughout the world's temperate and tropical climates, green-backed herons are most at home perched on rocks, roots, or tree branches, waiting for prey to come within reach. They seldom wade but will stand quietly in very shallow water in protected areas. When potential prey comes within range, this little predator may begin to stretch out, with neck and legs extended, in anticipation of striking and may even leap from its perch into the water.

This species is one of the few in the heron/egret group that prospers in years when no pronounced drydown occurs. It is easily understood why the green-backed heron is a poor competitor in a shallow, dry season pond full of other much larger wading birds. However, what this little heron lacks in physical ability, is compensated by innovation: some individuals learn to attract prey with bits of food. This phenomenon has often been observed where fish food is available, such as in zoos and outdoor public aquariums. The birds will find a piece of food, drop it in the water, and catch the fish that arrive to eat it. In the wild, individuals may use berries or other objects to attract curious small fish. This learned, specialized skill helps the green-backed heron feed in locations where fishing might otherwise be poor and where there is little competition.

Black-Crowned and Yellow-Crowned Night-Herons

While most other herons will feed at night, the night-herons, which are common in southern Florida, are specialists at nocturnal feeding and are seldom seen in daytime feeding groups of the other wading birds. Both species have large eyes, which are an incredible deep red in adults. Both feed heavily on crabs, crayfish, and other large aquatic invertebrates, but will take fish without hesitation. In a display of creativity similar to the green-backed heron, a juvenile black-crowned night-heron was observed catching a damselfly, which it did not eat. After killing the insect with a pinch, it placed the damselfly on the water surface and then slowly backed away six or eight inches to watch motionlessly. A minute passed as the damselfly drifted about foot, when the heron carefully retrieved the floating insect, minced it slightly, and repeated the process—several times. No fish approached, but the bird obviously appeared to be fishing, a behavior not normally associated with this species.[131a] Better known are the black-crowned night-heron's occasional infamous raids of nesting colonies to feed on the chicks of other wading birds.

Wading Bird Rookeries[171a]

Most of the wading birds nest in large colonies, typically in trees on small islands such as mangrove islands in Florida Bay or tree islands in the Everglades, but white ibis have been known to nest in sawgrass. The social nesting aggregations are called *rookeries* and usually contain a mixture of species, with great egrets perhaps nesting close to tricolored herons and white ibis. Some birds fly great distances from their rookery to foraging locations. Wood storks, unusually adept at soaring on rising air currents (thermals), often commute as far as 35 miles,[171] and a maximum of 80 miles has been observed.[173] Great egrets normally fly relatively shorter distances, and the majority travel less than five miles; however, small numbers do fly more than 15 miles and occasionally as far as 25 miles. Most white ibis commute less than six miles, and while much longer flights have been recorded, they are usually by birds from failing (starved) colonies.[13a]

Rookeries commonly contain anhingas, cormorants, and brown pelicans in addition to wading birds. A few wading birds, such as green-backed herons and reddish egrets, do not join rookeries but instead nest in dispersed locations. Outside of the breeding season, most of these species and other wading birds disperse, using shifting sites for roosting.

Most rookeries of the Everglades were formerly located in the southern part of the ecosystem, in Everglades National Park. Rookery Branch, located at the head of the Shark River and once home to the largest of the region's rookeries, is now abandoned. In the 1930s, it harbored as many as 200,000 birds, mostly white ibis. Many of the birds that formerly nested in the park have moved north into the Water Conservation Areas, where newly established rookeries have been unstable. Corkscrew Swamp, the most famous existing rookery in southern Florida, is located in the Big Cypress Swamp. It is almost entirely dominated by wood storks. Chapter 18 includes a discussion of the major problems of rookeries.[73, 74, 92, 118, 206]

Threatened and Endangered Species

Several Everglades birds are listed as *threatened* or *endangered* under Florida and/or federal regulations.[259a] In addition to these categories, Florida regulations also contain a status of lesser jeopardy called *species of special concern*, which has been given to several wading birds such as the little blue and tricolored herons. Prominent issues in management of the Everglades have focused on three endangered birds: the wood stork (discussed earlier), the snail kite, and the Cape Sable seaside sparrow.

Snail Kite[16, 16a, 111]

The snail kite (formerly called the Florida Everglade kite) is related to hawks and eagles and is about the same size and shape as the commonly seen red-shouldered hawk. It differs from the hawk in that it is generally dark brown (female) or slate black (male), with white at the base of the tail. The beak of the snail kite is narrow with a pronounced hook. This adaptation allows it to eat its primary food, the applesnail, which it snatches in its talons from the water while hovering. Snail kites seldom take other prey, but they have been known to catch small turtles and crayfish when snails are scarce.

The specialized diet renders the snail kite extremely habitat specific to shallow, long-hydroperiod marshes that contain sufficient applesnail populations, where the snails are visible and the water surface is not obstructed by dense vegetation. Sloughs with scattered emergent vegetation and long-hydroperiod wet prairies are optimal habitats, provided that trees are locally available for perching. Snails are most vulnerable while breathing; they are usually caught while on vegetation within about four inches of the surface and are probably safe from predation much deeper than that.

It was formerly thought that snail kites were not apt to relocate in search of suitable habitat when local conditions deteriorated, but tracking studies have shown that individuals often move great distances throughout their range in Florida, which extends from the southern Everglades northward to the St. Johns marshes east of Orlando. However, the Florida population does not migrate and is therefore separate from other populations of the species living in Cuba and Central and South America.

Cape Sable Seaside Sparrow[111, 199a]

The Cape Sable seaside sparrow is a small, handsomely striped, olive, grey, brown, and white non-migratory bird that characteristically lives in short-hydroperiod, freshwater or brackish marshes, where it builds nests woven of grasses and suspended in dense grass tussocks. Almost the entire range of this endemic subspecies of the seaside sparrow is the southern Everglades and Big Cypress Swamp areas within Everglades National Park. It feeds on insects and spiders and leads a secretive, highly sedentary lifestyle that is vulnerable to fire and water management. This bird attracted attention when a monitored population was estimated to have decreased by 95 percent between the mid-1950s and the mid-1970s. The dusky seaside sparrow of the Cape Canaveral area, a closely related subspecies, was officially recognized as extinct in January 1991.

A Contest of Beauty

Details of southern Florida's avifauna can go on endlessly; however, a contest of beauty presents a fitting finale. The most striking of the water birds (and not formerly mentioned) is the purple gallinule. This bird has a general shape similar

FIGURE 17.5
An American swallow-tailed kite carries dinner—a small snake—over the Everglades. (Photo by R. Hamer.)

to a small chicken, but it exhibits truly unbelievable colors, which are most enhanced in breeding plumage. The bright red beak is tipped in yellow. The forehead is light blue, and the remainder of the head and the entire breast are a deep purple-blue. The back is an iridescent dark green. The sight of one of these long-toed birds walking across lily pads in the sunlight is breathtaking. The most striking land bird is a winter resident, the male painted bunting. Sparrow-sized with somewhat secretive habits, it is not easily found, but is surprisingly worth the search. With a fire-red breast and rump, bright green back, and rich, dark blue head, it seems almost mythical. How such color can survive in the wild stretches the imagination. However, based on form and grace alone, the species that stands out is the elegantly graceful American swallow-tailed kite (Figure 17.5),[260] a relative of the hawk, which is a spring and summer breeding resident of southern Florida.

Part IV
Environmental Impacts

18

Man and the Everglades

T he foregoing pages have dealt primarily with natural history, confining most references to human impact on the Everglades ecosystem* to the introduction. Thus, it may appear that the future of the ecosystem will be its fate relative to a continued rise in sea level, a topic addressed numerous times. From the time of the first human footprints in Florida, nearly 14,000 years ago, the rise in sea level was dramatic until about 6000 years ago, when it began slowing to a very gradual rate, which has been maintained over the last 3000 years. During the period of most rapid rise, it would have come up about five feet per century. During the last 3000 years, however, rising sea level has been only a minor cause of environmental change.[251, 252a] Modern man's impact on the Everglades has made rising sea level seem remote and unimportant in comparison. Nevertheless, recent understanding of human-induced global warming indicates that rising seas may begin to play a significant role in the future of the Everglades, in addition to the continuation of other, more immediate and ongoing environmental problems for which impacts on southern Florida's natural systems are evident everywhere. These problems range from taking (collecting) plants and wildlife, to the introduction of exotic plants and animals, to extensive alterations of the natural hydrology of the region.

---

* The impact on the Everglades of the original Native Americans of the region is little known. When Europeans arrived, an estimated 20,000 Native Americans inhabited southern Florida, including the Calusa (generally to the west and north of the Everglades) and the Tequesta (to the east and south). Few lived in the Everglades interior. These original populations dwindled to an apparent end in 1763. The modern Seminoles and Miccosukees in the Everglades region came from Alabama and Georgia beginning in the early 1800s.[28, 198, 228]

Obviously, the various Native American groups had some environmental impact due to hunting and the use of fire (see Chapter 3), but their effects on the natural environment must have been small relative to the impacts that have occurred since the late 1800s. The historical presence of Native Americans can be regarded as part of the natural ecosystem.

Specimen Collecting

Collectors have taken their toll on plants and wildlife. Many of the more unusual orchids and ferns of the region are now all but extinct.[39, 41] The intricately colored Florida tree snails, once found in a myriad of varieties throughout the tropical hardwood hammocks of the area, have been collected so copiously for their beautiful shells that numerous varieties no longer exist. In fact, tree snail collectors in the past were even known to burn hammocks in order to facilitate collecting the indigenous variety of snail, damaging both the hammocks as well as the snail populations.[14] Reptile collectors can also damage populations, and the rarity of the large and beautiful indigo snake (an important predator of rattlesnakes) is partially their doing.[159] Collecting—much of which is now illegal—will continue to be a problem for wildlife managers in the future.

Off-Road Vehicles[56a, 57, 207a]

A variety of off-road vehicles (ORVs) have evolved for traveling in wetlands where the water is too shallow or obstructed for conventional boats and soils are too soft for cars or motorcycles. ORVs have greatly enhanced man's ability to enter remote areas where few would otherwise venture. Four general categories of ORVs are used in the variable terrain of the Big Cypress Swamp, including the small, three-wheeled all-terrain cycle (ATC) and four-wheeled all-terrain vehicle (ATV); a plethora of improvised versions of the "swamp buggy" (typically four-wheel drive vehicles with large aircraft or tractor tires); various tracked vehicles, either half-track with front steering tires or full-track; and airboats. In the open, flat, more flooded Everglades terrain, the use of airboats (flat-bottom boats driven by an aircraft or automotive engine and pusher propeller) far surpasses other types of vehicles, with small numbers of tracked vehicles also used.

The direct impact of ORVs includes soil rutting, damage to vegetation, spread of introduced species, noise, aesthetic (visual) alteration of wilderness, specimen collecting (especially in remote areas), and intrusive interference with wildlife. Secondary impacts include alteration of water flow and interruption of fire, with attendant functional changes to the ecosystem. In the Big Cypress, where swamp buggies and tracked vehicles have been used frequently primarily for hunting, soil rutting has been extensive in some areas, and major "mud highways" have developed, especially near entry points. Areas of marl soils are particularly damaged because the ruts are very slippery when wet. Subsequent ORVs avoid the earlier ruts, making additional tracks. One such area became a quarter of a mile wide where almost all vegetation was eliminated. Aside from some major ORV trails and their substantial visual impacts (especially from aerial views), damage in the Big Cypress Swamp is considered to be small, although the impact to wildlife due to human intrusion is little understood. In the Everglades Water Conservation Areas, tracked vehicles have substantially damaged tree islands.[68a, 211]

The impact of airboats in the Everglades has been considered relatively minor. Their operation is limited to official use in Everglades National Park and to strictly regulated use in the Loxahatchee National Wildlife Refuge, but there is extensive commercial and private recreational use in the Water Conservation Areas and other wetlands that lie between them. Relative to other ORVs, airboats cause the least damage to soils and vegetation, but repeated use of the same trails results in uprooting vegetation and subsequent soil displacement, producing channels. Some areas are strewn with such trails (Figure 18.1). The hydrological and biological impacts of soil disturbance and altered flow patterns (reducing sheet flow) are not well understood. However, cattail establishment is commonly associated with areas of heavy airboat traffic, and cattails often invade abandoned trails, even those lightly traveled.[131a]

The impact of airboats on wildlife is also little known, but the effect of

FIGURE 18.1
Airboat trails in Northeast Shark Slough. The smallest trails are the result of a single pass. Studies have shown such impacts to be negligible, but trails used repeatedly (such as the large diagonal example) become watercourse alternatives to sheet flow. This and the impact of harassment of wildlife by airboats have received little study.

harassment may by problematic. Noise effects would seem obvious, as most airboats are excessively loud, but noise by itself has been extensively studied (mostly with respect to aircraft) and results indicate that wildlife easily habituate to constant or repeated noise and sudden, surprise noises cause only startle reactions. Physical intrusion and noise combined are far more damaging; because airboats often travel 40 to 60 miles per hour (and even higher speeds are sometimes achieved), the possibility of significant adverse wildlife reactions seems inevitable. Further evaluation is warranted.[60a, 81a, 81b, 84b]

Introduced Species

Of the introduced biota, perhaps the greatest threat to the Everglades is a tree originating from Australia. It is often called *melaleuca* after its scientific name (*Melaleuca quinquenervia*), but it is also known by several common names: punk tree, cajeput, and broad-leaved paperbark (its name in Australia) (Figure 18.2). Its sensitivity to freezing weather restricts it to the southern half of the Florida peninsula; it grows in uplands and wetlands, where it efficiently invades sawgrass marshes as well as cypress swamps, especially where water levels have been reduced by drainage. It forms dense forests which exclude essentially all native vegetation and which transpire more water than marsh vegetation, thus (theoretically) tending to dry up marshes it has invaded. The spread of melaleuca is enhanced by its extreme tolerance of fire. Huge melaleuca forests are spreading into the Everglades from both coasts, where they have been upsetting the natural food chains of the native wetlands. Whether or not melaleuca can be prevented from overtaking Everglades National Park remains uncertain, but its effects outside the park are already far-reaching. Some ecologists consider melaleuca to be the single most serious threat to the future integrity of the Everglades ecosystem.[12, 50, 103]

Melaleuca may be the most ominous exotic pest plant, but it is not the only one. Second to melaleuca in its potential severity in the Everglades ecosystem is a small tree from South America called Brazilian pepper (*Schinus terebinthifolius*), sometimes misleadingly called Florida holly. Originally introduced as an ornamental, its seeds have been widely spread by birds. It has invaded a number of the plant communities of the Everglades region, especially pinelands and a wide variety of coastal lowland habitats including shallow mangrove swamps. Particularly troublesome is its ability to form a closed forest in shallow coastal marsh habitats. Extensive areas of the western part of Everglades National Park are now Brazilian pepper forests and are effectively ruined as feeding habitat for water birds. Still other problems have been caused by the Australian pine (*Casuarina equisetifolia*) and downy-myrtle (*Rhodomyrtus tomentosa*). The latter is a shrub which has spread rapidly into a variety of habitats near Naples, Florida and may soon make headlines. Well over 200 species of introduced plants have now invaded Everglades National Park, and the list of troublesome species will grow as more exotics escape cultivation.[15, 52]

FIGURE 18.2
An invaded sawgrass marsh. Melaleuca (at left) and Australian pine (at right) team up against
the native plant community, but melaleuca will win in this case, as evidenced by the
numerous seedlings in the area. An average of 25 years passes from the time seedlings first
appear in a sawgrass marsh until the area becomes a closed forest of melaleuca. This area
is just east of U.S. Highway 27 at the southern edge of Broward County.

In addition to introduced plants, the Everglades also supports numerous
introduced animals, the most prominent of which are fishes. The walking catfish,
which was much publicized in the 1970s, has spread throughout the Everglades,
where it has become abundant only in man-altered habitats. This species does not
appear to present an adverse impact on the Everglades ecosystem and it is often
seen being eaten by a variety of water birds and by the Florida gar. On the other
hand, the blue tilapia may be a problem. It becomes too large to be eaten by water
birds and is a herbivore, substantially altering the aquatic plant community.
Beginning in the late 1970s, it became abundant at the Anhinga Trail in Everglades National Park, where it (probably with help from the Mayan cichlid and
spotted tilapia) likely eliminated growths of naiad (*Najas* sp.), a submerged

aquatic plant. Naiad had been an important cover for smaller fish, as well as a food supply supporting a large population of golden shiners which, in turn, were important prey for numerous water birds (see front cover). This and other impacts of exotic fishes have received little study, but may be of substantial importance to wading birds and other ecosystem functions.[133, 137]

Hydrology and Land Use[92, 127a]

Of all the changes to South Florida's natural systems, tampering with the water of the vast expanses of wetlands has had the greatest impact on the integrity of the Everglades ecosystem. The general history of alterations to the Everglades hydrology, from the Kissimmee River through Lake Okeechobee and the Everglades, is outlined in the introduction to this book by Marjory Stoneman Douglas. Early drainage efforts sought to reduce water levels in particular to develop agriculture in marsh and swamp lands, with little regard for conserving water for irrigation and potable uses. Fresh water was thought to be inexhaustible, and flood control efforts focused on directing interior waters through canals to coastal tidal waters as quickly as feasible. Within a few decades, however, uncontrolled drainage brought many problems, notably endangering the water supply for the human population, causing uncontrollable soil fires with attendant choking smoke, and in fact not preventing floods in very wet years such as 1947. Therefore, flood control progressed to a higher level of technology on a massive scale: water management, as represented by the birth of the Central and Southern Florida Project for Flood Control and Other Purposes (C&SF Project), authorized by Congress in 1948 (Figure 18.3). With implementation beginning in the mid-1950s and the main features completed by the mid-1960s, the subsequent water management has been highly beneficial to many of man's interests—agriculture, water supply, and flood control—but not to wildlife.*

The C&SF Project had three main components. First, it established a perimeter levee through the eastern portion of the Everglades, blocking sheet flow so that lands farther east would be protected from direct Everglades flooding. The levee was about 100 miles long. It became the westward limit of agricultural, residential, and other land development for the lower east coast from West Palm Beach to Homestead. Only a few areas were subsequently developed west of it, most notably the 8.5-square-mile residential area of Dade County south of the Tamiami Trail, northwest of Homestead. The perimeter levee severed the eastern 16 percent of the Everglades from its interior (Figure 18.4).

Second, the C&SF Project designated a large area of the northern Everglades, south of Lake Okeechobee, to be managed for agriculture. Only a portion near the lake had previously been developed, leaving much room for agricultural expan-

---

* For historic perspectives, see References 17, 29, 127a, and 211. To be fair, it should be understood that there was no intent to ruin an ecosystem; rather, the intent was to harness and use land, and the CS&F Project was generally perceived to be the salvation of the Everglades.

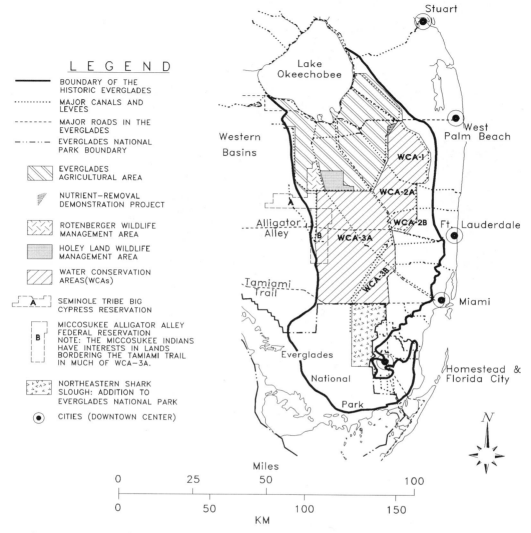

FIGURE 18.3

Map of the modern, subdivided Everglades, showing the major features of the C&SF Project.

sion in what was a vast, nearly unbroken expanse of sawgrass. Except for two large tracts—the Rotenberger and Holey Land (so named for its bomb craters!) Wildlife Management Areas—all of this area eventually was used for agriculture, primarily sugarcane. Called the Everglades Agricultural Area (EAA), it encompassed about 27 percent of the historic Everglades (Figure 18.5) and was a major factor in the economic justification of the C&SF Project.

Third, water conservation became the primary designated use for most of the remaining Everglades between the EAA and Everglades National Park, limited on the east by the eastern perimeter levee and on the west by an incomplete levee

FIGURE 18.4
Historic Everglades. Looking south-southeast from over the north leg of the aptly named Sawgrass Expressway (not visible) at a residential subdivision in northern Broward County. Development here is typical of areas east of the eastern perimeter levee (of WCA-2A in this case), involving removal of surficial muck and excavation of the underlying sand and limestone (often requiring blasting) in order to make lakes and waterways that provide fill for land areas and water storage and conveyance for flood protection. In some areas, lake depths exceed 50 feet due to the need to obtain sufficient fill for land development.

bordering the Big Cypress Swamp. The area was divided into three units (essentially wetland impoundments) called Water Conservation Areas (WCAs). The northernmost (WCA-1), also designated for wildlife management, eventually became the Arthur R. Marshall Loxahatchee National Wildlife Refuge and included most of the Hillsborough Lakes area of the Everglades, noted for its abundance of tree islands and sloughs. WCA-2 and WCA-3 (the largest) soon became problematic in water management and were subdivided into A and B units. The WCAs are separated from one another and from Everglades National Park by levees and are regulated by interconnecting canals and water control structures. Huge pump stations move water into or out of the WCAs, giving protection to the surrounding lands by receiving excess water at times of impending flood and by storing and selectively releasing water to compensate for drought conditions (Figure 18.6). They also deliver water to Everglades National Park via control structures along the Tamiami Trail (Figure 18.7). The WCAs encompass 32 percent of the historic Everglades.

FIGURE 18.5
Historic Everglades. A view of endless fields of sugarcane in the Everglades Agricultural Area southeast of Lake Okeechobee. The only sense of topographic relief is the illusion created by a "jog" in the system of land sections; the area is actually *flat*! Cleared, unplanted fields reveal the region's black muck soil of sawgrass origin.

Everglades National Park, established in 1947, contained only part of the southern Everglades, originally about 21 percent of the historic freshwater eco-system. Recently, another 4 percent has been added from an area west of the eastern perimeter levee, but south of WCA-3B, known as the Northeastern Shark Slough. This area had been overdrained but undeveloped. Its recent addition brings about 25 percent of the historic freshwater Everglades into Everglades National Park.

The C&SF Project divided the Everglades into pieces and had a profound effect on human existence in southeastern Florida, making land development safe and inviting. It controlled nature, but at the same time left huge areas in a natural state. What went wrong?

Decline of Wildlife

The story can best be told by referring to two contrasting case histories: the alligator and the wading birds (herons, egrets and their relatives), particularly the species that regularly use the interior freshwater wetlands.

FIGURE 18.6
Regional water management and the Everglades. The structure over the canal is the Acme Improvement District Pump Station No. 2. WCA-1 (Arthur R. Marshall Loxahatchee National Wildlife Refuge) is in the foreground, bordered by the L-40 rim canal and levee. Following the intent of the 1948 C&SF Project (see text), this pump station typifies regional water management, in this case for private lands. It helps control water levels in 19,000 acres of residential and agricultural land in Palm Beach County, using WCA-1 as a reservoir where excess waters can be pumped to prevent flooding (via a culvert through the levee) or where irrigation water can be withdrawn. Such private districts are authorized under Chapter 298, Florida Statutes, but must conform to requirements of the South Florida Water Management District. Most of these "two-ninety-eight districts" connect indirectly to the water conservation areas via the major drainage canals and control structures operated by the South Florida Water Management District. (Photo by T. Lodge.)

Protection of the alligator has been a success story, but it took the removal of economic incentives in the sale of alligator hides (by blocking the final sale of products) to ensure the alligator's dramatic recovery from projected extinction. So it was, in times past, with certain of the wading birds. Some of the species were hunted close to extinction during the final years of the 19th century, because their fancy plumes were then fashionable. Laws protecting these birds were partially

FIGURE 18.7
Looking northward over the Tamiami Trail where its crosses Water Control Structure 12C, one of four main delivery points in the Shark River Slough where water from WCA-3A (background) enters Everglades National Park (foreground). The smaller parallel road is the original Tamiami Trail completed in 1928, which was replaced by the newer road with the construction of the levee and control structures of WCA-3 in the early 1960s. (Photo by T. Lodge.)

effective, but it was the success of a public awareness campaign during the early years of this century to make the fashion socially unacceptable that finally removed the economic incentive for plume hunters. Populations of most of these birds then rebounded during the next few decades, to a high point for this century in about 1935.[200]

Since the mid-1930s, and with measurably increased intensity after the 1950s, a farther reaching dilemma has affected the wading birds of southern Florida. A general understanding of how these birds used the Everglades is instructive at this point. The historic ecosystem was a contiguous vastness, with a mosaic of plant communities and a cycle of wet and dry seasons. During each dry season, the feeding activity of these birds followed the drying trend as it progressed through the wetlands. In typical years, flocks followed the prime conditions of drydown, where prey were concentrated in low areas of the Everglades mosaic of plant communities, on a day-by-day basis. This progression of a "drying front" removed the guesswork for the birds. For an assured food supply for themselves and their nestlings, the birds returned to the location where they had been the previous day, with a minor correction for the single-day increment in the drying front location. Because the expanses of wetlands of the historic ecosystem were so huge and contained such a range of hydroperiods, a long breeding season was almost guaranteed. Conditions would be right somewhere, and thus the great wading bird rookeries of the region prospered.[13a, 47a, 172]

Most of the historic rookeries were located at the southern end of the Everglades,* along the upper tidal creeks of the great mangrove forests. Numerous conditions favored that location. It provided ready access to a greater range of hydroperiods and habitats than any other location in the entire Everglades ecosystem. The very long hydroperiod of the Shark River Slough was confined to a relatively narrow path (less than eight miles wide) through the southern Everglades, which was flanked by a gradation of freshwater habitats extending to the short-hydroperiod (six to seven months) marl prairie with abundant solution holes, the rocky glades. As typical dry seasons progressed, drying fronts moved toward the rookeries from all directions—inward toward the edges of the Shark River Slough and generally southward through the northern and central expanses of Everglades—but normally a large, shallow, freshwater pool was left above the buttonwood embankment at the end of the dry season. In addition, the great flow of fresh water through the historic ecosystem supported a highly productive estuarine mangrove zone, which provided habitat and prey base alternatives to the freshwater Everglades throughout the year on regular tidal cycles. Thus, within short range, the birds could take advantage of rich habitat diversity and probably the best fish production in the region, but this was all based on freshwater flows from the north—dependable flows that changed only gradually in the transition from wet to dry season.[13a, 47a, 73, 136, 248]

The initial drainage of the Everglades diminished the amounts of water that moved southward through the Shark River Slough, Taylor Slough, and into the estuaries.[216] The wading birds and their southern Everglades rookeries persisted through this stage, however. It has only been since the 1950s, after the implementation of the C&SF Project, that their numbers have diminished greatly. The nesting population in the rookeries in the area before the first insult—plume hunting—is poorly documented, but estimates run as high as 2.5 million. Plume hunters reduced this number drastically, but recovery from that time to the mid-1930s returned nesting populations to impressive numbers, which indicates that the ecosystem was still intact.

Since that time, populations have again fallen, slowly through the 1950s and then dramatically, as shown in Table 18.1. The historic giant rookeries of the southern Everglades are essentially gone. New rookeries which developed in the

---

* Much smaller rookeries occurred elsewhere in the Everglades and along the shores of Lake Okeechobee, but the available records indicate that they were minor and unstable compared to the giant southern Everglades rookeries. Also, comparable to the southern Everglades location, large rookeries at the Usumacinta-Grijalva system are located at the freshwater marsh–mangrove interface.[170]

TABLE 18.1
Estimated Breeding Population of Selected Water Birds Since the 1930s
in Everglades National Park and Adjacent Cental and Southern Everglades Habitats[a]

Species	1931–1946	1974–1981	1982–1989
great egret	6,500	6,500	4,200
small herons[b]	25,000	8,000	2,500
white ibis	200,000	29,000	12,500
wood stork	6,500	2,600	750

[a] Source: Ogden[170] and modified from Ogden[171a] (Table 22.2 therein).

[b] Small heron counts are almost entirely snowy egrets and tricolored herons, species that utilize rookeries and would have been taken by plume hunters.

WCAs north of Everglades National Park, associated with the only remaining long-hydroperiod marshes, were the result of impoundment in southern portions of the WCAs. However, these colonies are smaller and appear unstable in their frequent shifts and nesting failures. Reduced productivity of the estuarine area, loss of much of the short-hydroperiod rocky glades of the southern Everglades, and loss of the long hydroperiod in the lower Shark River Slough correlate with the demise of the southern Everglades rookeries, and a large general impact has resulted from compartmentalization of the ecosystem. Formerly, drying fronts progressed great distances, but with the C&SF Project, they became interrupted by levees. Conditions on opposite sides of levees are invariably different, either already dry or too deep, which forces the birds to search for another area of prey concentration, which may not exist. Pulsed releases of water have also been damaging. Use of the WCAs for flood control and unnatural releases of water from WCA-3A to Everglades National Park have reversed drying fronts, eliminating prey concentrations and causing nesting failures. It has become apparent that the compartmentalized, managed, and reduced ecosystem supports successful nesting of wading birds only under very special conditions, namely rapid, uninterrupted drydown following a productive wet season, such as the favorable conditions in the 1991–92 winter/spring season. Under these particular conditions, enough prey is concentrated to facilitate successful but short wading bird nesting seasons.[13a, 73, 171a, 200]

Many performance modifications to the C&SF Project, including both structural and water delivery modifications, have attempted to reverse the downward trend in wading bird populations, but with little apparent success. Oddly enough, the alligator has persisted through these problems and has endured less harm than the wading birds, largely because poaching has been eliminated. However, the important function of the alligator in assisting the wading birds has been minimized. Previously, alligator holes and trails were extremely important to these birds, both as sites where prey collected during low water and where some prey survived to repopulate adjacent marshes in the rainy season. Accentuated drying of the Everglades in April and May, during the normal alligator courtship period, has reduced gator nesting. Also, due to water management practices early in the wet season, too much water has often flooded the nests, killing the eggs.[149a] As a result, alligators are now more confined to man-made habitats. The huge service historically provided by "rural," "homesteading" alligators throughout the Everglades marshes has been undone. Canals suit alligators adequately, but most are too deep and too steep-sided for effective use by wading birds, especially wood storks. Although the birds line up along the edges of canals during the dry season, the prey availability is low compared to drying gator holes or other low areas in the marsh where food is trapped by receding water. Thus, wading birds are a roving workforce, dependent upon an abundant, harvestable, but moving resource that has become reduced and unpredictable. In contrast, the alligator exploits a wide variety of local resources and even beneficially modifies its environment; when all else fails, it retreats to aestivate through the hard times. This age-old reptile—largely unchanged since the time of the dinosaurs—continues to survive.

Water Quality Impacts[46a]

Added to the foregoing hydrologic problems is another complication: the discharge of nutrient-laden water from the EAA. The earnest efforts of water managers to keep agricultural water out of Lake Okeechobee, by diverting it instead into Water Conservation Areas after 1979, is thought to have been responsible for an extensive proliferation of cattails.* Areas dominated by cattails do not favor the development of beneficial food supplies for wading birds, and thus even more wetland is effectively removed from their utilization. However, areas of dense sawgrass are also known to provide poor habitat for Everglades wildlife, thus compounding the issue. Most experts agree that nutrient impacts on the Everglades ecosystem is a growing problem, but is of small magnitude compared to the history of hydrologic impacts of altered water quantities and flows.[5]

Because of low background levels, phosphorus is the key consideration in nutrient enrichment of the Everglades. Considerable effort has been devoted to developing the technology to remove phosphorus from waters potentially entering the Everglades Protection Area (defined as the WCAs and Everglades Na-

---

* Water quality has become a major legal controversy, initiated on October 27, 1988, when the federal government sued the state of Florida (specifically the Department of Environmental Regulation and the South Florida Water Management District) for violations of Florida's own standards in waters entering the Arthur R. Marshall Loxahatchee National Wildlife Refuge and waters directed to Everglades National Park. While the lawsuit[124] focused attention on the issue of degraded waters, untold dollars that might have been spent on correcting the problems have been diverted to pay for legal action.[211, 212] The reader should be aware that cattail infestations can result from factors other than nutrients, such as increased hydroperiod and water depth, which would favor cattails over sawgrass.[3, 46a] Infestation as the result of increased nutrients must be proven on a case-by-case basis, which involves examining a variety of evidence.

tional Park). The South Florida Water Management District has constructed a demonstration project for wetland treatment which consists of 4000 acres abutting the northwest side of the Arthur R. Marshall Loxahatchee National Wildlife Refuge. The formation of peat soil is the intended long-term means of phosphorus removal. The goal is to reduce the average influent concentration of phosphorus from 175 ppb to about 50 ppb; if successful, as much as 35,000 additional acres for stormwater treatment may be constructed to treat EAA waters.[85a, 110, 212, 213b]

Mercury

An additional result of tampering with the hydrology of the Everglades may be the recently unveiled mystery of high levels of mercury in some of the fauna of the region. Mercury, familiar as a silvery metallic liquid, forms chemical compounds of various toxicities. Of particular concern in aquatic environments is methylmercury, the highly toxic, organically bound form, which readily enters living organisms. It may cause mutations and abnormal growths, as well as neurologic disorders which may lead to death at higher concentrations. Methylmercury is apparently being concentrated in the Everglades food web, with predators containing the highest levels. Even in remote areas of the Everglades, largemouth bass contain so much mercury that authorities have had to warn fishermen of the dangers of eating them, as well as certain other species. Animals that feed on aquatic life, such as raccoons and alligators, have also accumulated mercury.[112, 254]

In mid-1989, following growing concern over the mercury issue, a Florida panther was found dead in the Everglades. Liver samples from this individual contained 110 ppm of mercury, which indicated that mercury poisoning may have been the cause of death. This particular panther frequented the Shark River Slough, where deer (which have low mercury levels) were scarce, and it is known that raccoons and alligators formed a large part of its diet.[109, 134, 137a] Then, in mid-1991, the last two female panthers inhabiting Everglades National Park died. Their mercury levels were lower than the earlier case, but mercury poisoning was still regarded as a probable contributor to the death of at least one of these individuals.[169]

The origin of the mercury is not known at this time. Agriculture was initially suspected because of the use of mercury compounds in some agricultural applications. However, the distribution of mercury levels in soil, water, and biological samples has not supported this notion. One theory is that it occurs naturally in the peat soils of the Everglades. Through the last 5000 years, while the Everglades soils were forming, small amounts of mercury became chemically trapped there. Then, with the recent drainage and subsequent oxidation of the peat soils, the mercury became concentrated and moved into the food web. However, another probable contributor is airborne deposition from such sources as burning sugarcane fields, incineration of medical and other wastes, paint applications, and fossil-fueled power plants.[63, 109, 112, 134] Whatever the source, a possible inverse relationship has been observed between environmental phosphorus levels and mercury content in fish tissue, which hints that higher phosphorus levels inhibit or mitigate mercury uptake, while lower levels allow for more effective bioconcentration in the food chain.[185a]

The nature and extent of the mercury dilemma is under investigation. It is not yet known, for example, whether mercury is contributing to the demise of wading bird populations in the Everglades.[75a] Mercury appears to be of critical importance, and much more research is needed in order to understand its implications, as well as to determine any possible remediation.

Eustrongylidosis[75a, 218a]

Eustrongylidosis is a parasitic disease of fish-eating birds. This cumbersome name is derived from the scientific name of the causative worm, *Eustrongylides ignotus*. It has been responsible for the death of many wading birds, and an epidemic was first documented in portions of the Everglades in the mid-1980s. *Eustrongylides* is a nematode, or roundworm, and its life cycle begins when its eggs hatch into larvae that infect small, bottom-dwelling, freshwater oligochaetes, which are worms related to earthworms. When the infected oligochaetes are eaten by any of numerous fishes, the larval roundworms migrate into the fish tissue, where they remain coiled in a cyst. If an infected fish is eaten by a wading bird, the encysted worms mobilize and migrate through the stomach wall into other body organs, where they become reproductive adults. The migration and the activity of the adult worms cause substantial damage and kill most infected nestlings. Adult birds are more tolerant, but serve the parasite by distributing its eggs, which enter the bird's digestive tract and are released in feces, completing the parasite's life cycle.

Various diseases are common among wildlife species, and occasional epidemics affecting one or more species occur just as they do in human populations. However, the recent eustrongylidosis epidemic may be an impact of land development. Infected fish are most often found where waters have been polluted by various wastes, notably in residential areas, probably because the oligochaetes (required in the parasite's life cycle) are more abundant in enriched sediments. Like the mercury issue, much more research is needed to understand the implications of this disease on wading bird populations.

Florida Bay[152a]

Florida Bay was historically renowned for its excellent commercial and sport fishing and for its abundance of wading birds. Beginning in the 1970s, however, declines were observed, which greatly intensified from the mid-1980s into the early 1990s. About 15 square miles of seagrass beds died in the western half of the bay between 1987 and 1990, and about 90 square miles were damaged.[195] By 1992, over 50 square miles had died.[13] Sporadic planktonic algal blooms and extensive losses of wading birds, fish, shrimp, sponges, and even mangroves of the islands of the bay were documented.[18, 243]

Initial investigations focused on hypersalinity from diverted freshwater flows. A historic maximum salinity of 75 ppt (over twice the salinity of sea water) was recorded in May 1990, and a particularly interesting study involving a type of coral found in the bay added credibility. The coral was found to record the presence of humic compounds of freshwater origin in its growth bands, which indicated decreased freshwater inputs to the bay beginning in 1912; this corre-lated with the initial Everglades drainage works.[216] There was much public criticism of reduced flows through Taylor Slough and adjacent eastern portions of Everglades National Park, where fresh water had been diverted by drainage projects and where the history (later including the "Frog Pond" agricultural area) was particularly troublesome to conservationists. In the mid-1960s, a major drain-age canal, the C-111 canal (Figure 18.8), was constructed near the eastern bound-ary of Everglades National Park.* Together with earlier canals and other drainage modifications, C-111 further interrupted and diverted flows from Taylor Slough— water that would have naturally moved through the eastern area of the park into Florida Bay through a broad area of sheet flow.

In spite of the circumstantial connection to altered freshwater inputs, an ad hoc committee of scientists who evaluated the Florida Bay situation concluded that the disruption of freshwater flows contributed to hypersaline-related prob-

---

* When C-111 was under construction (1964–66), citizens observed that it was larger and deeper than comparable drainage canals and that a drawbridge was being provided for the U.S. 1 highway crossing. The design had been intended—without the public participation required after 1969 by the National Environmental Policy Act (NEPA)—to serve as a drain-age canal *and* to allow barge traffic to carry rocket assemblies between the Kennedy Space Center and a proposed Aerojet General Corporation testing facility located near the main entrance to Everglades National Park. The testing project was never funded, but public outrage over the construction of a canal that might seriously affect the waters of a national park was enormous (even the National Park Service voiced a strenuous but unheeded objection) and precipitated a lawsuit by the National Audubon Society against the U.S. Army Corps of Engineers. An even larger concern than the drainage aspect of the project was saltwater intrusion. It was resolved that an earthen plug would remain near the mouth of the canal (later replaced by a water control structure) to prevent the inland movement of salt water.[2, 29, 147] The water control structure remains today, as does the alternate name for C-111, the Aerojet Canal. (The previous reputation of the Corps of Engineers now stands in contrast to its role as a steward of wetland protection under the Clean Water Act and as the lead federal agency in Everglades restoration.)

Even the National Park Service has been responsible for controversial canal projects, notably the Buttonwood Canal in Everglades National Park. When the canal was completed in 1957, it allowed salt water from Florida Bay to enter Coot and eastern Whitewater bays, which were rapidly and drastically altered to saline from their historic freshwater/brackish condition.[39, 243] Subsequent installation of a barrier in the canal at the Flamingo marina in July 1982 may have then aggravated problems developing in Florida Bay, although the intent to correct the historic environment of Coot and Whitewater bays was admirable.

FIGURE 18.8

The extreme southeastern Everglades, looking northwestward over Canal C-111, with the U.S. Highway 1 drawbridge at the center. The dam and water control structure (S-197) that prevent saltwater intrusion and regulate freshwater flow are at the lower left. The entire C-111 history has been controversial, but added drama occurred on August 15, 1988, when the dam was removed (for the fifth time since August 1981) for eight days as an emergency flood protection measure. The resulting freshwater discharge to Barnes Sound then killed marine life over a vast area.[211] Also note that the canal and highway cut across the north-south orientation of the tree islands, which indicates the direction of historic sheet flow. (Photo by T. Lodge.)

lems only in the northeastern portion of the bay. Based on the limited data available, the larger areas of the bay had not been affected by intensified hyper-salinity beyond historically recurring conditions. It was recognized that the deteriorated environment in much of Florida Bay may have resulted from a natural cause, namely the lack of a direct hit by a major storm since Hurricanes Donna (1960) and Betsy (1965) (Andrew's wrath passed mostly north of Florida Bay). The intervening period of relative quiet allowed natural shoaling and sediment accumulation, which restricted circulation and flushing and produced an increasingly stagnant condition that hampered many ecological functions. Also recognized was the fact that data on the bay were insufficient to determine whether water quality problems were related to man-induced pollution,[18] a highly controversial topic.

Others have considered that Florida Bay's problems may have been aggravated by the extreme drought of 1989–90 and by the original construction of the Overseas Railroad (the alignment now used by U.S. Highway 1 through the Keys) which cut off potentially important circulation to and from neighboring Barnes Sound and the Atlantic in the early 1900s. In spite of the complexities and unknowns, many scientists and the public are convinced that Florida Bay is the end-of-the-line recipient of the altered hydrology of the Everglades and have urged numerous ongoing water management changes for Taylor Slough and C-111 in order to correct deficiencies in freshwater flows.[213]

Solving the Problems: Everglades Restoration[47a]

Certainly something new must happen in order to reestablish the breeding populations, especially the rookeries within the southern Everglades. The region cannot simultaneously support large populations of wading birds during their nesting period and the conflicting demands of agriculture and urban development, given the existing water control system. Restoration must involve more than modifying the operation of the existing C&SF Project. The various tinkerings have shown that approach to be fruitless.[103a]

The basis for Everglades restoration must come from an in-depth understanding of how the ecosystem functioned prior to the impacts of drainage and land development. As a focus, the factors that made it a superlative environment for many species of wading birds must be understood. Improving the reproductive success of wading birds—within the context of natural landscapes—is probably the best encapsulation of an overall Everglades restoration goal. The high position of wading birds in the food chain makes them excellent indicators of a wide variety of wetland functions that result in the production and availability of their prey.[74, 171a]

Many important details of the historic Everglades are poorly known, with only scattered descriptions by early observers to establish the actual historic conditions. Especially instructive are general descriptions of the vegetation, the connection of Lake Okeechobee to the Everglades, and accounts of the locations of the major wading bird rookeries. Details of the historical hydrology and of water quality are lacking, but extensive alternative evidence has been developed and recently evaluated to supplement the lack of baseline data, which allows estimates of the historic ecosystem compared to its altered state today. Useful evidence has emerged from the following:

- Hydrologic modeling, which has predicted that the original water depths, hydroperiods, and flows in the natural ecosystem were far greater than today's conditions in the remaining Everglades and that the pattern of vegetation provided for gradual water level changes, unlike the rapid changes often caused by today's water management[66a]

- Water quality studies involving periphyton, sawgrass, and cattails, showing that the central and southern Everglades benefitted from low-nutrient conditions[22]

- Evaluations of changes in vegetation patterns through the last half century, confirming that altered hydrology has degraded the original and beneficial mosaic of plant communities[47, 256a]

- Soil studies, revealing past vegetation community patterns and the long history of fire[84a]

- Observations concerning the beneficial role of intense fire in Everglades landscapes in relation to wading bird success[73, 99a]

- Fish population studies, revealing which historic conditions likely provided a larger prey base for wading birds[134a]

- Observations on the changing role of the alligator in the Everglades[149a]

- Examination of the effects of reduced freshwater inputs to the estuarine and marine environments at the southern end of the Everglades[152a]

- Evaluation of factors that could explain the occurrence of the historic wading bird rookeries at the mangrove–Everglades interface and their recent occurrence in the WCAs[13a, 171a, 248]

All of these lines of evidence point to similar conclusions: most of the ecosystem has been denied water, reducing its depths and consequent flows to the southern estuaries and especially reducing hydroperiods through most of the system. As a result, sawgrass has spread into wet prairie and slough communities which were once too deep for its survival.[256a] Alligators have relocated from peripheral wetland areas, now too dry for their success, and moved into slough communities that were originally too wet to accommodate their life cycles. Salinities of the mangrove estuaries have increased, and productivity has noticeably fallen.

These studies also make it increasingly clear that the long-term cycles of drought and fire were important. Severe droughts allowed intense soil fires in the late dry season. Such fires did not burn the entire ecosystem, but instead created a patchwork of depressions in the Everglades landscape in which long-hydroperiod slough communities developed. Subsequent succession of plant communities led to wet prairies and finally to sawgrass. Thus, soil fire was the factor responsible for the mosaic pattern of different plant communities. The synergism of side-by-side deeper and shallower areas of the mosaic promoted the production of, and accessibility to, a large fish biomass, which benefitted wading birds. However, modern management of the Everglades has strived to perpetuate the status quo, using such techniques as frequent control burns to avert the possibility of catastrophic fire. Combined with reduced water, plant community succession began to erase the mosaic.[47]

The possible effects of extensive wet season flooding, which may have served to spread aquatic animal species throughout the landscape, into locations not reachable in normal or dry years, are not as well understood. Levees now divide the ecosystem, preventing widespread migration of aquatic life. Further, water management has incorporated weather prediction; emergency water releases are initiated days prior to the arrival of tropical weather systems, thus reducing the effect of natural flooding. There have been obvious benefits to man's residential and agricultural interests, but nature has probably suffered. The ability of management to separate the systems, imitating natural events in the natural system while protecting the entrenched human infrastructure, should be the challenge of the future.

Probably the least understood of any historic conditions, also true of today's Everglades, is the role of invertebrates and amphibians. Studies of these animal groups are needed to assist in formulating restoration strategies.

Restoration Issues

The path to Everglades restoration will be filled with unknowns and problems. For example, water movement through wetlands is affected by many factors, which makes its simulation by modeling very complex; yet the success of restoration will require particular seasonal conditions of hydroperiod, depth, and flow to the estuaries. Thus, it is improbable that a plan even of massive proportions could be devised which would reach the goals of ecosystem restoration in one effort. A successful plan must be defined and implemented in numerous phases, with refinements and corrections for any shortcomings left for later phases. Engineering features must be designed with flexibility in order to accommodate probable needs to change operation.

An issue of monumental stature will be the delivery of water from Lake Okeechobee to the Everglades, now accomplished by routing through canals and pumping into the WCAs. The possibility of a direct wetland corridor through the EAA has been raised,[255] and the major loss of existing agricultural and residential land that would be required has sparked tremendous concern among potentially affected agribusinesses and citizenry. Concern for the impact of Lake Okeechobee's degraded waters on the water quality of the Everglades is also involved.

Two complex and interrelated issues for Everglades restoration will be water discharges from the "western basins" (agricultural lands west of the EAA) and possible increased flooding of Indian lands. The western basins now discharge water southward through a canal system constructed through the Seminole Tribe Big Cypress Reservation and into reservation lands of the Miccosukee Indians, where it ends in WCA-3A. Water quality concerns and flooding associated with restoration will undoubtedly present new difficulties to an already controversial topic.

Conflicts between government agencies have arisen due to differing philosophies and legalities. For example, the National Park Service takes an ecosystem approach—to attain, or maintain, historically natural conditions. The Fish and Wildlife Service focuses on certain species, especially threatened and endangered species, and strives for adjustments that maximize benefits to them. This philosophical difference became problematic during agency assessment of proposed modifications for restoring flows through WCA-3B to the Northeastern Shark Slough. The proposal, viewed as ecosystem restoration by the National Park Service, was viewed by the Fish and Wildlife Service as possible jeopardy for the snail kite in WCA-3A, which would receive less water. The issue was resolved in mediation arranged by the National Audubon Society,[176] but requirements of the

Endangered Species Act will undoubtedly cause other conflicts with broad-spectrum ecosystem alterations.

An almost certain problem for Everglades restoration will be a tendency for public opinion to influence water and fire management in order to duplicate "typical" or "average" years all of the time. Few will want to face the perceived cruelties caused by excessive flooding (with attendant drowning of alligator nests, deer, bobcats, etc.) or severe drought with intense fires. It is now clear, however, that these longer-term cyclic rigors were of key importance to the Everglades ecosystem, in addition to the regular wet season/dry season annual cycle. The Everglades is not a region of constant tranquility, and attempting to maintain the status quo by duplicating a so-called average year will surely lose the battle.

It is obvious that restoration to mimic the original condition of the Everglades may directly involve little more than half of the original area, namely the Everglades Protection Area—Everglades National Park and the Water Conservation Areas. However, historic Everglades lands outside of the perimeter levees must make contributions to restoration goals involving water storage and wetland values. Additional land use restrictions will be highly controversial, with dangerous legal entanglements, but a few policies may be easily accepted and highly beneficial. For example, the historic Everglades terrain in Palm Beach, Broward, and Dade counties (much of which is still wetlands, although degraded) has been undergoing rapid development, and this trend will undoubtedly continue. The more recent wetland impacts in these areas have required mitigation (restoration, enhancement, or creation of compensating wetlands), but there has been no coordinating policy on how mitigation should relate to the overall function of the Everglades ecosystem. Coordination of future mitigation projects to benefit the Everglades restoration effort could significantly improve its outcome.[131b]

Finally, the overall restoration plan must have the ecosystem-wide cooperation of the many vested interests, both public and private. In particular, all government interests—local, state, and federal—must be coordinated in *one* direction. Separate, overlapping, and sometimes conflicting agendas are counterproductive, and local regulations that apply only to parts of the ecosystem (i.e., within political jurisdictions) are potentially detrimental to the needed coordination. To expand on the preceding mitigation example, future wetland impacts in the historic Everglades could be coordinated to produce a series of "mitigation banks," large mitigation projects constructed to collectively compensate for wetland impacts in various locations. Mitigation banks could be specifically designed to maximize production of prey for wading birds, and their annual drydowns could be artificially sequenced for wading bird needs. In critical years, certain species, such as the wood stork, could be specifically targeted. Such a system could dramatically outperform the disjunct, local mitigation projects now accomplished under a complex overlap of local, regional, state, and federal regulations which do not provide for cooperation across political boundaries. Clearly, one ultimate permitting authority should coordinate all wetland mitigation projects throughout the historic Everglades. Expertise from various levels is helpful, but the implications of wetland permitting to Everglades restoration are too important to be controlled without regard for overall ecosystem objectives.

Unless the numerous local governments with jurisdiction in the historic Everglades become part of a coordinated effort, they should not have wetland permitting authority. At least six counties currently regulate the historic Everglades, not to mention the Okeechobee–Kissimmee watershed.

Restoration Guidelines

The recent intense efforts to understand the Everglades in terms of needed restoration lead to the following objectives:

- Reduce losses of fresh water, currently released to the sea through canal systems, and retain amounts in the remaining Everglades to mimic the historically longer hydroperiods. Reestablishing contributing flows from the Okeechobee–Kissimmee watershed is part of this effort.

- Reestablish southerly sheet flow to the estuaries through Shark River and Taylor sloughs, which involves the removal of interior dikes and canals.

- Avoid damaging water management practices, such as artificially accentuated reversals of winter/spring drydown (which interrupt the food supply for wading birds) and excessive flooding (which destroys alligator nests in the wet season).

- Eradicate deleterious introduced vegetation and continue to improve controls on future introductions of both plants and animals.

- Continue the rising use of Best Management Practices and continue evaluation of stormwater treatment methods for reducing nutrients from agriculture and urban development in waters entering the Everglades. Waters within the Everglades must be forced to follow sheet flow, to provide additional treatment and to prevent "short circuiting" of nutrients through the system.

- Improve habitat management techniques to reestablish the historic Everglades mosaic pattern.

- Improve the "hydroperiod diversity" of the ecosystem, within the range from short hydroperiod (six to seven months) to nearly continuous inundation (but without artificially impounding water as in the southern portions of the existing WCAs). Special attention should be focused on evaluating and rehabilitating wetlands with very short hydroperiod (less than about five months) or with depths that do not reach at least a foot during the height of typical wet seasons.

- Coordinate wetland mitigation projects in the historic Everglades with restoration objectives.

- Evaluate the compatibility of recreational uses of the Everglades with restoration goals and modify as necessary.

- Continue research on mercury and wildlife diseases in the Everglades, including their origins, impacts, and possible mitigation.

- Set clear goals for restoration phases and monitor performance in order to provide guidance for succeeding phases.

The Short-Term Prognosis: The Next Few Decades

If the advance of melaleuca and Brazilian pepper cannot be checked, and in time they come to dominate the Everglades, there would be just reason to abandon the idea of a national park named *Everglades*. Control programs are in place for melaleuca at least and are purportedly slowing the spread. However, the work is very expensive because of its intensive labor requirements, use of herbicides, and transportation by helicopter. Obviously, a biological control (now under investigation) must be found soon.[146, 160, 235]

Provided that exotic trees are controlled, the native tropical vegetation of the region will continue to be of interest. Alligators and sport fishing will continue to be prominent attractions. The recovery of wading bird populations, however, will require large-scale restoration of Everglades hydrology. A greatly improved understanding of the ecosystem can provide fruitful guidance for restoration, and promising initiatives have developed.[63, 162a, 213a, 255] The momentum toward restoration is encouragingly strong, but without implementation *and* significant results, the Everglades will not regain any semblance of its once spectacular abundance of beautiful wading birds—the main reason that an earlier generation sought to protect part of the ecosystem as a national park.[146a] The forecast[14] made in 1938 by biologist Daniel Beard (later to become the first superintendent of Everglades National Park) was all too prophetic:

> Practically without exception, areas that have been turned over to the [National Park] Service as national parks have been of superlative value with existing features so outstanding that if the Service were able to merely retain the status quo, the job was a success. This will not be true of the Everglades National Park. The reasons for even considering the lower tip of Florida as a national park are 90 percent biological ones, and hence highly perishable. Primitive conditions have been changed by the hand of man, abundant wildlife resources exploited, woodland and prairie burned and reburned, water levels altered, and all the attendant, less obvious ecological conditions disturbed.
>
> Director Cammerer recently said, "I would much rather have a national park created that might not measure up to all everybody thinks of it at the present time, but which, 50 or 100 years from now, with all the protection we could give it, would have attained a natural condition comparable to primitive conditions..." If the National Park Service is prepared to follow the strategy thus expressed, the Everglades National Park seems justified. If it is not ready to do this, the writer wishes to state emphatically, the Everglades is *not* justified.

The Long-Term View

What is the future beyond the next several decades? No matter what is done in terms of ecosystem restoration, another threat will likely emerge: rising sea level. Projections based on sea level data recorded at locations along the east coast of the United States, including Florida, indicate a rate of rise of about one foot per century since 1930. In other words, it has already risen over six inches since then!

This rate is an abrupt increase compared to the rate of only about one foot per *thousand* years over the last 3000 years. At the present pace, it would still require five centuries to flood most of Everglades National Park in salt water. Even at this pace, however, extensive near-term impacts are to be expected. Foremost among these is that the huge mangrove swamps of the region will be rapidly decimated—hurricane by hurricane—and replacement swamps will not have time to mature before they too are overtaken. The advancing shoreline will have a narrower fringe composed of younger mangroves. Such a trend is now in progress, with extensive inland establishment of mangrove seedlings during the last few decades and many storm-damaged mangrove areas now too deep for seedling reestablishment.[252a]

The current opinion among a majority of scientists is that the world is on the course of an even more rapid inundation of coastal areas than observed since 1930. This prediction is based on the greenhouse effect, which is produced by rising atmospheric levels of carbon dioxide, methane, and synthetic chlorofluorocarbons.[83, 97, 215] Resulting warmer temperatures will induce faster melting of polar ice, most importantly the Antarctic ice cap. While controversy rages on global warming,* conservative predictions now estimate sea level rise at one-and-a-half feet per century. More radical predictions estimate a faster rate. We can only speculate what the actual rate will be, but the demise of the southern Everglades—Everglades National Park—may not be many generations in the future. Today's children will likely have to deal with serious problems of rising seas, in one walk of life or another, during adulthood.

The possibility of worldwide warming and the resulting increases in sea level may make the local problems of introduced species, water management, and land use in southern Florida seem small and solvable by comparison. One is swept by a feeling of helplessness amidst the need for international cooperation. This is not to say that the aggravating atmospheric problems cannot be solved; certainly there will be grave concerns among all nations that face coastal flooding as well as the possible adverse impacts on agriculture and fisheries. Clearly, Everglades restoration must proceed at several levels. For its own survival, mankind must

---

* Within this controversy, a well-founded minority predicts a return to a cooling trend with subsequent glaciation and a reduction of sea level based on the influence of the Milankovitch cycles, which should reduce solar energy received by the earth.[184] Whatever the future holds, there is an ongoing paradox: Florida has experienced some of the most severe cold weather in its history in spite of the recent warming trend. Before the 1980s, central Florida was the center of Florida's vast citrus industry. However, a series of winter freezes since 1980, culminating in the devastating Christmas week freeze of 1989, eliminated citrus from areas of the state where it had prospered for nearly a century. The 1989 freeze even damaged mangroves in the Florida Keys, a rare phenomenon. Can such weather be consistent with projections of a worldwide warming trend? It must be borne in mind that global warming means an increase in *average* temperature, mostly due to heating in the tropical and temperate zones of the *summer* hemisphere. With more heat energy available to drive entire weather systems, perhaps we can expect more movement of air masses, cold as well as warm. One uncomfortable outcome may be that more frequent and colder masses of Arctic air reach Florida in the winter (author's speculation).

unite to save the earth's remaining natural amenities, such as wilderness areas. Their value is not just to those who benefit as direct "users," such as tourists, campers, and wildlife photographers. Natural areas actually represent a measure of the integrity of the life support system that has made the earth fit for mankind in the first place. In the case of the Everglades, it would be a sad commentary for future generations to have to say, "At least we have some photographs."

References

1. Abrahamson, Warren G. and David C. Hartnett. 1990. Pine flatwoods and dry prairies. *In* Ronald L. Myers and John J. Ewel (editors). *Ecosystems of Florida*. University of Central Florida Press, Orlando, Florida.

2. Alexander, Taylor R. 1967. Effect of Hurricane Betsy on the southeastern Everglades. *Quarterly Journal of the Florida Academy of Sciences* 30(1):10–24.

3. Alexander, Taylor R. 1971. Sawgrass biology related to the future of the Everglades ecosystem. *Soil and Crop Science Proceedings* 31:72–74.

4. Alexander, Taylor R. 1974. Evidence of recent sea level rise derived from ecological studies on Key Largo, Florida. *In* P.J. Gleason (editor). *Environments of South Florida: Present and Past*. Memoir 2, Miami Geological Society, Miami, Florida.

5. Alexander, Taylor R. (personal communications). Professor of Botany Emeritus, Department of Biology, University of Miami, Coral Gables, Florida.

6. Alexander, Taylor R. and Alan G. Crook. 1984. Recent vegetational changes in southern Florida. *In* P.J. Gleason (editor). *Environments of South Florida: Present and Past II*. Memoir II, Miami Geological Society, Miami, Florida.

7. Ashton, Ray E., Jr. and Patricia Sawyer Ashton. 1988. *Handbook of Reptiles and Amphibians of Florida, Part One—The Snakes*. Windward Publishing, Miami, Florida.

8. Ashton, Ray E., Jr. and Patricia Sawyer Ashton. 1988. *Handbook of Reptiles and Amphibians of Florida, Part Three—The Amphibians*. Windward Publishing, Miami, Florida.

9. Ashton, Ray E., Jr. and Patricia Sawyer Ashton. 1991. *Handbook of Reptiles and Amphibians of Florida, Part Two—Lizards, Turtles & Crocodilians* (revised second edition). Windward Publishing, Miami, Florida.

10. Avery, George N. 1980. Plants of Everglades National Park, A Preliminary Checklist of Vascular Plants. Report T-574, U.S. National Park Service, South Florida Research Center, Everglades National Park, Homestead, Florida.

11. Avery, George N. and Lloyd L. Loope. 1980. Endemic Taxa in the Flora of South Florida. Report T-558, U.S. National Park Service, South Florida Research Center, Everglades National Park, Homestead, Florida.

12. Balciunas, Joseph K. and Ted D. Center. 1991. Biological control of *Melaleuca quinquenervia*: prospects and conflicts. *In* Ted D. Center, Robert F. Doren, Ronald H. Hofstetter, Ronald L. Myers, and Louis D. Whiteaker (editors). Proceedings of the Symposium on Exotic Pest Plants. National Park Service, U.S. Department of the Interior, Washington, D.C.

13. Bancroft, G. Thomas. 1993. Florida Bay—an endangered North American jewel. *The Florida Naturalist* 66(1):4–9.

13a. Bancroft, G. Thomas, Allan M. Strong, Richard J. Sawicki, Wayne Hoffman, and Susan D. Jewell. Relationships among wading bird foraging patterns, colony locations, and hydrology in the Everglades. *In* Steven M. Davis and John C. Ogden (editors). *Everglades: The Ecosystem and Its Restoration.* St. Lucie Press, Delray Beach, Florida.

14. Beard, Daniel B. 1938. Wildlife Reconnaissance, Everglades National Park Project. National Park Service, U.S. Department of the Interior, Washington, D.C.

15. Bennett, Fred D. and Dale H. Habeck. 1991. Brazilian peppertree—prospects for biological control in Florida. *In* Ted D. Center, Robert F. Doren, Ronald H. Hofstetter, Ronald L. Myers, and Louis D. Whiteaker (editors). Proceedings of the Symposium on Exotic Pest Plants. National Park Service, U.S. Department of the Interior, Washington, D.C.

16. Bennetts, Robert E. 1993. The snail kite, a wanderer and its habitat. *The Florida Naturalist* 66(1):12–15.

16a. Bennetts, Robert E., Michael W. Collopy, and James A. Rodgers, Jr. 1994. The snail kite in the Florida Everglades: a food specialist in a changing environment. *In* Steven M. Davis and John C. Ogden (editors). *Everglades: The Ecosystem and Its Restoration.* St. Lucie Press, Delray Beach, Florida.

17. Blake, N.M. 1980. *Land into Water — Water into Land: A History of Water Management in Florida.* University Presses of Florida, Tallahassee, Florida.

18. Boesch, Donald F., Neal E. Armstrong, Christopher F. D'Elia, Nancy G. Maynard, Hans W. Paerl, and Susan L. Williams. 1993. Deterioration of the Florida Bay Ecosystem: An Evaluation of the Scientific Evidence. Report dated September 15, 1993, to the Interagency Working Group on Florida Bay sponsored by the National Fish and Wildlife Foundation, the National Park Service, and the South Florida Water Management District.

19. Bosence, Daniel. 1989. Biogenic carbonate production in Florida Bay. *Bulletin of Marine Science* 44(1):419–433.

20. Briggs, John C. 1958. A list of Florida fishes and their distribution. *Bulletin of the Florida State Museum, Biological Sciences* 2(8):224–318.

21. Brooks, H.K. 1984. Lake Okeechobee. *In* P.J. Gleason (editor). *Environments of South Florida: Present and Past II.* Memoir II, Miami Geological Society, Miami, Florida.

22. Browder, Joan A., Patrick J. Gleason, and David R. Swift. 1994. Periphyton in the Everglades: spatial variation, environmental correlates, and ecological implications. *In* Steven M. Davis and John C. Ogden (editors). *Everglades: The Ecosystem and Its Restoration.* St. Lucie Press, Delray Beach, Florida.

23. Brown, Randall B., Earl L. Stone, and Victor W. Carlisle. 1990. Soils. *In* Ronald L. Myers and John J. Ewel (editors). *Ecosystems of Florida.* University of Central Florida Press, Orlando, Florida.

23a. Burt, William Henry and Richard Philip Grossenheider. 1964. *A Field Guide to the Mammals* (second edition). Houghton Mifflin, Boston, Massachusetts.

24. Cantrell, Richard W. (personal communications). Florida Department of Environmental Protection, Water Management Division, Bureau of Wetland Resource Management, Tallahassee, Florida.

25. Carmichael, Pete and Winston Williams. 1991. *Florida's Fabulous Reptiles and Amphibians.* World Publications, Tampa, Florida.

26. Carr, Archie and Coleman J. Goin. 1959. *Guide to the Reptiles, Amphibians and Fresh-water Fishes of Florida.* University of Florida Press, Gainesville, Florida.

27. Carr, Robert S. (personal communication). Archaeological & Historical Conservancy, Inc., Miami, Florida.

28. Carr, Robert S. and John G. Beriault. 1984. Prehistoric man in southern Florida. *In* P.J. Gleason (editor). *Environments of South Florida: Present and Past II.* Memoir II, Miami Geological Society, Miami, Florida.

29. Carter, Luther J. 1974. *The Florida Experience.* Resources for the Future, Inc., The Johns Hopkins University Press.

30. Chávez, Miguel (editor). 1988. *Ecología y Conservación del Delta de los ríos Usumacinta y Grijalva* (Memorias). Instituto Nacional de Investigacíon sobre Recursos Bióticos (INIREB)—División Regional Tabasco, 720 pp. (This hard-to-find document is the proceedings of a symposium held in Tabasco, Mexico, in 1987. It includes papers on the Everglades. A copy is available in the South Florida Water Management District library, West Palm Beach, Florida.)

31. Chen, Ellen and John F. Gerber. 1990. Climate. *In* Ronald L. Myers and John J. Ewel (editors). *Ecosystems of Florida.* University of Central Florida Press, Orlando, Florida.

32. Clausen, Carl J. and Cesare Emiliani. 1979. Little Salt Spring preserver of the past. *Sea Frontiers* 25(5):258–265.

33. Cohen, Arthur D. 1984. Evidence of fires in the ancient Everglades and coastal swamps of southern Florida. *In* P.J. Gleason (editor). *Environments of South Florida: Present and Past II.* Memoir II, Miami Geological Society, Miami, Florida.

34. Cohen, Arthur D. and William Spackman. 1984. The petrology of peats from the Everglades and coastal swamps of southern Florida. *In* P.J. Gleason (editor). *Environments of South Florida: Present and Past II.* Memoir II, Miami Geological Society, Miami, Florida.

35. Comp, G.S. and W. Seaman, Jr. 1985. Estuarine habitat and fishery resources of Florida. *In* William Seaman, Jr. (editor). *Florida Aquatic Habitat and Fishery Resources.* Florida Chapter of the American Fisheries Society, Kissimmee, Florida.

36. Craighead, Frank C., Sr. 1963. *Orchids and Other Air Plants of the Everglades National Park.* Everglades Natural History Association, University of Miami Press, Coral Gables, Florida.

37. Craighead, Frank C., Sr. 1964. Land, mangroves, and hurricanes. *The Fairchild Tropical Garden Bulletin* 19(4):1–28.

38. Craighead, Frank C., Sr. 1968. The role of the alligator in shaping plant communities and maintaining wildlife in the southern Everglades. *The Florida Naturalist* 41:2–7; 69–74.

39. Craighead, Frank C., Sr. 1971. *The Trees of Southern Florida.* Volume 1: The Natural Environments and Their Succession. University of Miami Press, Coral Gables, Florida.

40. Craighead, Frank C., Sr. 1971. Is man destroying South Florida? *In* William Ross McCluney (editor). *The Environmental Destruction of South Florida.* University of Miami Press, Coral Gables, Florida.

41. Craighead, Frank C., Sr. 1984. Hammocks of south Florida. *In* P.J. Gleason (editor). *Environments of South Florida: Present and Past II.* Memoir II, Miami Geological Society, Miami, Florida.

42. D'Avanzo, Charlene. 1990. Long-term evaluation of wetland creation projects. *In* Jon A. Kusler and Mary E. Kentula (editors). *Wetland Creation and Restoration: The Status of the Science.* Island Press, Washington, D.C.

43. Davies, Thomas D. and Arthur D. Cohen. 1989. Composition and significance of the peat deposits of Florida Bay. *Bulletin of Marine Science* 44(1):387–398.

44. Davis, Gary E., 1992. Assessment of Hurricane Andrew Impacts on Natural and Archaeological Resources of Big Cypress National Preserve, Biscayne National Park, and Everglades National Park (Executive Summary). U.S. National Park Service, University of California, Davis, Cooperative Park Study Unit, California.

45. Davis, Gary E. and Jon W. Dodrill. 1989. Recreational fishery and population dynamics of spiny lobsters, *Panulirus argus,* in Florida Bay, Everglades National Park, 1977–1980. *Bulletin of Marine Science* 44(1):78–88.

46. Davis, J.H., Jr. 1943. The Natural Features of Southern Florida, Especially the Vegetation, and the Everglades. Geological Bulletin No. 25, The Florida Geological Survey, Tallahassee, Florida.

46a. Davis, Steven M. 1994. Phosphorus inputs and vegetation sensitivity in the Everglades. *In* Steven M. Davis and John C. Ogden (editors). *Everglades: The Ecosystem and Its Restoration.* St. Lucie Press, Delray Beach, Florida.

47. Davis, Steven M., Lance H. Gunderson, Winifred A. Park, John R. Richardson, and Jennifer E. Mattson. 1994. Landscape dimension, composition, and function in a chang-

ing Everglades ecosystem. *In* Steven M. Davis and John C. Ogden (editors). *Everglades: The Ecosystem and Its Restoration*. St. Lucie Press, Delray Beach, Florida.

47a. Davis, Steven M. and John C. Ogden. 1994. Toward ecosystem restoration. *In* Steven M. Davis and John C. Ogden (editors). *Everglades: The Ecosystem and Its Restoration*. St. Lucie Press, Delray Beach, Florida.

47b. DeAngelis, Donald L. 1994. *In* Steven M. Davis and John C. Ogden (editors). *Everglades: The Ecosystem and Its Restoration*. St. Lucie Press, Delray Beach, Florida.

48. Dennison, Mark S. and James F. Berry. 1993. *Wetlands: Guide to Science, Law and Technology*. Noyes Publications, Park Ridge, New Jersey.

49. Dewar, Heather. 1993. Nature still reeling from storm. *The Miami Herald*, Tuesday, August 24, 1993, final edition, pp. 1A and 11A.

50. Diamond, Craig, Darrell Davis, and Don C. Schmitz. 1991. Economic impact statement: the addition of *Melaleuca quinquenervia* to the Florida Prohibited Aquatic Plant List. *In* Ted D. Center, Robert F. Doren, Ronald H. Hofstetter, Ronald L. Myers, and Louis D. Whiteaker (editors). Proceedings of the Symposium on Exotic Pest Plants. U.S. Department of the Interior, National Park Service, Washington, D.C.

51. Dineen, J. Walter. 1984. Fishes of the Everglades. *In* P.J. Gleason (editor). *Environments of South Florida: Present and Past II*. Memoir II, Miami Geological Society, Miami, Florida.

52. Doren, Robert F., Louis D. Whiteaker, and Anne Marie LaRosa. 1991. Evaluation of fire as a management tool for controlling *Schinus terebinthifolius* as secondary successional growth on abandoned agricultural land. *Environmental Management* 15(1):121–129.

53. Dorschner, John. 1993. Grasping Andrew's full fury. *The Miami Herald*, Thursday, July 22, 1993, final edition, pp. 1A and 15A.

54. Douglas, Marjory Stoneman. 1988. *The Everglades: River of Grass* (revised edition). Pineapple Press, Sarasota, Florida.

55. Douglas, Marjory Stoneman (personal communications). Coconut Grove, Florida, 1989 to 1993.

56. Duever, Michael J. 1984. Environmental factors controlling plant communities of the Big Cypress Swamp. *In* P.J. Gleason (editor). *Environments of South Florida: Present and Past II*. Memoir II, Miami Geological Society, Miami, Florida.

56a. Duever, Michael J., John E. Carlson, and Lawrence A Riopelle. 1981. Off-Road Vehicles and Their Impacts in the Big Cypress National Preserve. Report T-614, National Park Service, South Florida Research Center, Everglades National Park, Homestead, Florida.

57. Duever, Michael J., John E. Carlson, John F. Meeder, Linda C. Duever, Lance H. Gunderson, Lawrence A. Riopelle, Taylor R. Alexander, Ronald L. Myers, and Daniel P. Spangler. 1986. The Big Cypress National Preserve. National Audubon Society Research Report No. 8, National Audubon Society, New York, New York.

57a. Duever, M.J., J.F. Meeder, L.C. Meeder, and J.M. McCollom. 1994. The climate of south Florida and its role in shaping the Everglades ecosystem. *In* Steven M. Davis and John C. Ogden (editors). *Everglades: The Ecosystem and Its Restoration*. St. Lucie Press, Delray Beach, Florida.

57b. Dunkle, Sidney W. 1989. *Dragonflies of the Florida Peninsula, Bermuda, and the Bahamas*. Scientific Publishers, Gainesville, Florida.

58. Duplaix, Nicole. 1990. South Florida water: paying the price. *National Geographic* 178(1): 89–113.

59. Egler, Frank E. 1952. Southeast saline Everglades vegetation in Florida, and its management. *Vegetatio Acta Botanica III* (Fasc. 4–5):213–265.

60. Eklund, A.M., W.F. Loftus, and M. Covi. (in preparation). Population Dynamics of the Freshwater Prawn (*Palaemonetes paludosus*). U.S. National Park Service, South Florida Research Center, Everglades National Park, Homestead, Florida.

60a. Ellis, David H., Catherine H. Ellis, and David P. Mindell. 1991. Raptor responses to low-level jet aircraft and sonic booms. *Environmental Pollution* 74:53–83.

61. Enos, Paul. 1989. Islands in the bay—a key habitat of Florida Bay. *Bulletin of Marine Science* 44(1):365–386.

62. Erwin, Kevin L. 1990. Freshwater marsh creation and restoration in the southeast. *In* Jon A. Kusler and Mary E. Kentula (editors). *Wetland Creation and Restoration: The Status of the Science.* Island Press, Washington, D.C.

63. Everglades Coalition. 1993. The Greater Everglades Ecosystem Restoration Plan. Everglades Coalition, Washington, D.C.

64. Ewel, Katherine C. 1990. Swamps. *In* Ronald L. Myers and John J. Ewel (editors). *Ecosystems of Florida.* University of Central Florida Press, Orlando, Florida.

65. Fairbridge, Rhodes W. 1984. The Holocene sea-level record in South Florida. *In* P.J. Gleason (editor). *Environments of South Florida: Present and Past II.* Memoir II, Miami Geological Society, Miami, Florida.

66. Faulkner, Barry M. and Albert V. Applegate. 1986. Hydrocarbon exploration evaluation of the Pulley Ridge area, offshore South Florida Basin. *Gulf Coast Association of Geological Societies Transactions* 36:83–95.

66a. Fennema, Robert J., Calvin J. Neidrauer, Robert A. Johnson, Thomas K. MacVicar, and William A. Perkins. 1994. A computer model to simulate natural Everglades hydrology. *In* Steven M. Davis and John C. Ogden (editors). *Everglades: The Ecosystem and Its Restoration.* St. Lucie Press, Delray Beach, Florida.

67. Fernald, Edward A. (editor). 1981. *Atlas of Florida.* The Florida State University Foundation, Tallahassee, Florida.

68. Fernald, Edward A. and Donald J. Patton (editors). 1984. *Water Resources Atlas of Florida.* Florida State University, Tallahassee, Florida.

68a. Florida Department of Transportation. 1991. Determination of Effectiveness of Animal Crossing Structures on I-75/Alligator Alley in Reducing Animal/Auto Collisions: Quarterly Performance Report. December 21, 1990–March 15, 1991.

69. Fogarty, Michael J. 1984. The ecology of the Everglades alligator. *In* P.J. Gleason (editor). *Environments of South Florida: Present and Past II.* Memoir II, Miami Geological Society, Miami, Florida.

70. Forthman, C.A. 1973. The Effects of Prescribed Burning on Sawgrass, *Cladium jamaicense,* in South Florida. Masters thesis, University of Miami, Coral Gables, Florida.

71. Franz, Richard (editor). 1982. *Volume Six: Invertebrates. In* Peter C.H. Pritchard (series editor). *Rare and Endangered Biota of Florida.* University Presses of Florida, Gainesville, Florida.

72. Franz, Richard and Shelly E. Franz. 1990. A review of the Florida crayfish fauna, with comments on nomenclature, distribution, and conservation. *Florida Scientist* 53(4): 286–296.

73. Frederick, Peter C. 1993. Wading Bird Nesting Success Studies in Water Conservation Areas of the Everglades. Final report to South Florida Water Management District, West Palm Beach, Florida.

74. Frederick, Peter C. and Michael W. Collopy. 1988. Reproductive Ecology of Wading Birds in Relation to Water Conditions in the Florida Everglades. Technical Report No. 30, Florida Cooperative Fish and Wildlife Research Unit, School of Forestry Research and Conservation, University of Florida.

75. Frederick, Peter C. and William F. Loftus. 1993. Responses of marsh fishes and breeding wading birds to low temperatures: a possible behavioral link between predator and prey. *Estuaries* 16(2):216–222.

75a. Frederick, Peter C. and Marilyn G. Spalding. 1994. Factors affecting reproductive success of wading birds (Ciconiiformes) in the Everglades ecosystem. *In* Steven M. Davis and John C. Ogden (editors). *Everglades: The Ecosystem and Its Restoration.* St. Lucie Press, Delray Beach, Florida.

76. Gentry, Cecil R. 1984. Hurricanes in South Florida. *In* P.J. Gleason (editor). *Environments of South Florida: Present and Past II.* Memoir II, Miami Geological Society, Miami, Florida.

77. George, Jean Craighead. 1972. Everglades Wildguide. U.S. Department of the Interior, National Park Service, Office of Publications, Washington, D.C.

78. Gifford, John. 1911. *The Everglades and Other Essays Relating to Southern Florida*. Everglade Land Sales Co., Kansas City, Missouri.

79. Gifford, John C. 1945. *Living by the Land*. Parker Art Printing Association, Coral Gables, Florida.

80. Gilbert, Carter R. (editor). 1992. *Volume II: Fishes. In* Ray E. Ashton, Jr. (series editor). *Rare and Endangered Biota of Florida*. University Presses of Florida, Gainesville, Florida.

81. Gilbert, Carter R. and Reeve M. Bailey. 1972. Systematics and Zoogeography of the American Cyprinid Fish *Notropis (Opsopoeodus) emiliae*. Number 644, Occasional Papers of the Museum of Zoology, University of Michigan.

81a. Gladwin, Douglas N., Duane A. Asherin, and Karen M. Manci. 1988. Effects of Aircraft Noise and Sonic Booms on Fish and Wildlife: Results of a Survey of U.S. Fish and Wildlife Service Endangered Species and Ecological Services Field Offices, Refuges, Hatcheries, and Research Centers (NERC 88/30). U.S. Fish and Wildlife Service National Ecology Research Center, Fort Collins, Colorado.

81b. Gladwin, Douglas N., Karen M. Manci, and Rita Villella. 1988. Effects of Aircraft Noise and Sonic Booms on Domestic Animals and Wildlife: Bibliographic Abstracts (NERC-88/32). U.S. Fish and Wildlife Service National Ecology Research Center, Fort Collins, Colorado.

82. Glantz, Michael H. 1984. Floods, fires, and famine: is El Niño to blame? *Oceanus* 27(2): 14–19.

83. Glantz, Michael H. 1990. Does history have a future? Forecasting climate change effects on fisheries by analogy. *Fisheries* 15(6):39–44.

84. Gleason, Patrick J., H. Kelly Brooks, Arthur D. Cohen, Robert Goodrick, William G. Smith, William Spackman, and Peter Stone. 1984. The environmental significance of holocene sediments from the Everglades and saline tidal plain. *In* P.J. Gleason (editor). *Environments of South Florida: Present and Past II*. Memoir II, Miami Geological Society, Miami, Florida.

84a. Gleason, Patrick J. and Peter Stone. 1994. Age, origin, and landscape evolution of the Everglades peatland. *In* Steven M. Davis and John C. Ogden (editors). *Everglades: The Ecosystem and Its Restoration*. St. Lucie Press, Delray Beach, Florida.

84b. Glegg, Stewart A.L. and J.R. Yon. 1989. Determination of Noise Source Height of Vehicles on Florida Roads and Highways. Report No. FHWA/FL/DOT/MO-89-382, Department of Ocean Engineering, Florida Atlantic University, submitted to the Florida Department of Transportation, Environmental Research & Training, Tallahassee, Florida.

85. Godfrey, Robert K. and Jean W. Wooten. 1979 and 1981. *Aquatic and Wetland Plants of the Southeastern United States* (Volumes 1 and 2). University of Georgia Press, Athens Georgia.

85a. Goforth, Gary, James Best Jackson, and Larry Fink. 1994. Restoring the Everglades. *Civil Engineering* 64(3):52–55.

86. Goodrick, Robert L. 1984. The wet prairies of the northern Everglades. *In* P.J. Gleason (editor). *Environments of South Florida: Present and Past II*. Memoir II, Miami Geological Society, Miami, Florida.

87. Gore, Rick. 1993. Andrew aftermath. *National Geographic* 183(4):2–37.

88. Gosner, Kenneth L. 1979. *A Field Guide to the Atlantic Seashore*. Houghton Mifflin, Boston, Massachusetts.

89. Grant, Peter R. 1986. *Ecology and Evolution of Darwin's Finches*. Princeton University Press, Princeton, New Jersey.

90. Grant, Verne. 1963. *The Origin of Adaptations*. Columbia University Press, New York, New York.

91. Gunderson, Lance H. 1989. Historical hydropatterns in wetland communities of Everglades National Park. *In* R.R. Sharitz and J. W. Gibbons (editors). *Freshwater Wetlands and*

Wildlife. DOE Symposium Series No. 61, United States Department of Energy, Office of Scientific and Technical Information, Oak Ridge, Tennessee.

91a. Gunderson, Lance H. 1994. Vegetation of the Everglades: determinants of community composition. *In* Steven M. Davis and John C. Ogden (editors). *Everglades: The Ecosystem and Its Restoration*. St. Lucie Press, Delray Beach, Florida.

92. Gunderson, Lance H. and William F. Loftus. 1993. The Everglades. *In* William H. Martin, Stephen G. Boyce, and Arthur C. Echternacht (editors). *Biodiversity of the Southeastern United States/Lowland Terrestrial Communities*. John Wiley & Sons, New York, New York.

92a. Gunderson, L.H. and J.R. Snyder. 1994. Fire patterns in the southern Everglades. *In* Steven M. Davis and John C. Ogden (editors). *Everglades: The Ecosystem and Its Restoration*. St. Lucie Press, Delray Beach, Florida.

93. Hancock, James and James Kushlan. 1984. *The Herons Handbook*. Harper & Row, New York, New York.

94. Harper, R.M. 1927. Natural Resources of Southern Florida. 18th Annual Report, Florida Geological Survey. Tallahassee, Florida.

95. Heald, E.J., W.E. Odum, and D.C. Tabb. 1984. Mangroves in the estuarine food chain. *In* P.J. Gleason (editor). *Environments of South Florida: Present and Past II*. Memoir II, Miami Geological Society, Miami, Florida.

96. Higer, Aaron L. and Milton C. Kolipinski. 1988. Changes in vegetation in Shark River Slough, Everglades National Park, 1940–1964. *In* Miguel Chávez Miguel (editor). *Ecología y Conservación del Delta de los ríos Usumacinta y Grijalva* (Memorias). Instituto Nacional de Investigacíon sobre Recursos Bióticos (INIREB)—División Regional Tabasco, pp. 217–230 (see Reference 30).

97. Hileman, Bette. 1992. Web of interactions makes it difficult to untangle global warming data. *Chemical & Engineering News* 70(17):7–19.

98. Hobbs, H., Jr. 1942. The crayfishes of Florida. *University of Florida Biological Science Series* 3(2):v+1–179.

99. Hobbs, H.H., III. (in press). *Trophic Relationships of North American Freshwater Crayfishes and Shrimps*. Milwaukee Public Museum, Milwaukee, Wisconsin.

99a. Hoffman, Wayne, G. Thomas Bancroft, and Richard J. Sawicki. 1994. Foraging habitat of wading birds in the Water Conservation Areas of the Everglades. *In* Steven M. Davis and John C. Ogden (editors). *Everglades: The Ecosystem and Its Restoration*. St. Lucie Press, Delray Beach, Florida.

100. Hoffmeister, John Edward. 1974. *Land from the Sea*. University of Miami Press, Coral Gables, Florida.

101. Hofstetter, Ronald H. 1984. The effect of fire on the pineland and sawgrass communities of southern Florida. *In* P.J. Gleason (editor). *Environments of South Florida: Present and Past II*. Memoir II, Miami Geological Society, Miami, Florida.

102. Hofstetter, Ronald H. 1988. Fire as a management tool in wetlands preservation: Florida Everglades. *In* Miguel Chávez Miguel (editor). *Ecología y Conservación del Delta de los ríos Usumacinta y Grijalva* (Memorias). Instituto Nacional de Investigacíon sobre Recursos Bióticos (INIREB)—División Regional Tabasco, pp. 245–257 (see Reference 30).

103. Hofstetter, Ronald H. 1991. The current status of *Melaleuca quinquenervia* in southern Florida. *In* Ted D. Center, Robert F. Doren, Ronald H. Hofstetter, Ronald L. Myers, and Louis D. Whiteaker (editors). Proceedings of the Symposium on Exotic Pest Plants. U.S. Department of the Interior, National Park Service, Washington, D.C.

103a. Holling, C.S., Lance H. Gunderson, and Carl J. Walters. 1994. The structure and dynamics of the Everglades system: guidelines for ecosystem restoration. *In* Steven M. Davis and John C. Ogden (editors). *Everglades: The Ecosystem and Its Restoration*. St. Lucie Press, Delray Beach, Florida.

104. Howe, William H. (editor). 1975. *The Butterflies of North America*. Doubleday & Company, Garden City, New York.

105. Humphrey, Stephen R. (editor). 1992. *Volume I: Mammals. In* Ray E. Ashton, Jr. (series

editor). *Rare and Endangered Biota of Florida*. University Presses of Florida, Gainesville, Florida.

106. Hunt, Burton P. 1952. Food relationships between Florida spotted gar and other organisms in the Tamiami Canal, Dade County, Florida. *Transactions of the American Fisheries Society* 82:14–33.

106a. Jaap, Walter C. and Pamela Hallock. 1990. Coral reefs. *In* Ronald L. Myers and John J. Ewel (editors). *Ecosystems of Florida*. University of Central Florida Press, Orlando, Florida.

107. Johnson, Lamar. 1974. *Beyond the Fourth Generation*. The University Presses of Florida, Gainesville, Florida.

108. Jones, A.C., S.A Berkeley, J.A. Bohnsack, S.A. Bortone, D.K. Camp, G.H. Darcy, J.C. Davis, K.D. Haddad, M.Y. Hedgepeth, E.W. Irby, Jr., W.C. Jaap, F.S. Kennedy, Jr., W.G. Lyons, E.L. Nakamura, T.H. Perkins, J.K. Reed, K.A. Steidinger, J.T. Tilmant, and R.O. Williams. 1985. Ocean habitat and fishery resources of Florida. *In* William Seaman, Jr. (editor). *Florida Aquatic Habitat and Fishery Resources*. Florida Chapter of the American Fisheries Society, Kissimmee, Florida.

109. Jordan, Dennis. 1990. Mercury contamination: another threat to the Florida panther. *Endangered Species Technical Bulletin* 15(2):1, 6.

110. Kadlec, Robert H. and Susan Newman. 1992. Phosphorus Removal in Wetland Treatment Areas. Technical report prepared for the South Florida Water Management District, West Palm Beach, Florida.

111. Kale, Herbert W., II (editor). 1978. *Volume Two: Birds. In* Peter C.H. Pritchard (series editor). *Rare and Endangered Biota of Florida*. University Presses of Florida, Gainesville, Florida.

112. KBN. 1992. Mercury Emissions to the Atmosphere in Florida (Final Report). Prepared for the Florida Department of Environmental Regulation by KBN Engineering and Applied Sciences, Inc., Gainesville, Florida.

113. Klitgord, Kim D., Peter Popenoe, and Hans Schouten. 1984. Florida: a Jurassic transform plate boundary. *Journal of Geophysical Research*. 89(B9):7753–7772.

114. Kreitman, Abe and Leslie A. Wedderburn. 1984. Hydrology of South Florida. *In* P.J. Gleason (editor). *Environments of South Florida: Present and Past II*. Memoir II, Miami Geological Society, Miami, Florida.

115. Kushlan, James. A. 1988. Impact of water management on wildlife in the Florida Everglades. *In* Miguel Chávez Miguel (editor). *Ecología y Conservación del Delta de los ríos Usumacinta y Grijalva* (Memorias). Instituto Nacional de Investigacíon sobre Recursos Bióticos (INIREB)—División Regional Tabasco, pp. 231–243 (see Reference 30).

116. Kushlan, James. A. 1990. Freshwater marshes. *In* Ronald L. Myers and John J. Ewel (editors). *Ecosystems of Florida*. University of Central Florida Press, Orlando, Florida.

117. Kushlan, James. A. and Thomas E. Lodge. 1974. Ecological and distributional notes on the freshwater fish of southern Florida. *Florida Scientist* 37(2):110–128.

118. Kushlan, James. A. and Deborah A. White. 1977. Nesting wading bird populations in southern Florida. *Florida Scientist* 40(1):65–72.

119. Lakela, Olga and Robert W. Long. 1976. *Ferns of Florida*. Banyan Books, Miami, Florida.

120. Lawson, Jack. 1989. Florida can't take ozone, greenhouse worries lightly. *Florida Environments*. Florida Environments Publishing, High Springs, Florida.

121. Layne, James N. (editor). 1978. *Volume One: Mammals. In* Peter C.H. Pritchard (series editor). *Rare and Endangered Biota of Florida*. University Presses of Florida, Gainesville, Florida.

122. Layne, James N. 1984. The land mammals of southern Florida. *In* P.J. Gleason (editor). *Environments of South Florida: Present and Past II*. Memoir II, Miami Geological Society, Miami, Florida.

123. Lazell, James D., Jr. 1989. *Wildlife of the Florida Keys: A Natural History*. Island Press, Washington, D.C.

124. Lehtinen, Dexter W. 1990. United States of America, plaintiff, vs. South Florida Water

Management District; John R. Wodraska, Executive Director, South Florida Water Management District; Florida Department of Environmental Regulation; and Dale Twachtmann, Secretary, Florida Department of Environmental Regulation, defendants. Second Amended Complaint. U.S. District Court Southern District of Florida, Case No. 88-1886-CIV-HOEVELER (signed January 4, 1990).

125. Levi, Herbert W., Lorna R. Levi, and Herbert S. Zim. 1968. *Spiders and Their Kin*. Golden Press, New York, New York.

126. Lewis, Roy R., III. (personal communication). Lewis Environmental Services, Inc., Tampa, Florida.

127. Lewis, R.R., III, R.G. Gilmore, Jr., D.W. Crewz, and W.E. Odum. 1985. Mangrove habitat and fishery resources of Florida. *In* William Seaman, Jr. (editor). *Florida Aquatic Habitat and Fishery Resources*. Florida Chapter of the American Fisheries Society, Kissimmee, Florida.

127a. Light, Stephen S. and J. Walter Dineen. 1994. Water control in the Everglades: a historical perspective. *In* Steven M. Davis and John C. Ogden (editors). *Everglades: The Ecosystem and Its Restoration*. St. Lucie Press, Delray Beach, Florida.

128. Little, Elbert L. 1980. *The Audubon Society Field Guide to North American Trees, Eastern Region*. Alfred A. Knopf, New York, New York.

129. Livingston, Robert J. 1990. Inshore marine habitats. *In* Ronald L. Myers and John J. Ewel (editors). *Ecosystems of Florida*. University of Central Florida Press, Orlando, Florida.

130. Lloyd, Jacqueline M. 1991. 1988 and 1989 Florida Petroleum Production and Exploration. Florida Geological Survey, Information Circular No. 107, Florida Department of Natural Resources.

131. Lodge, T.E. 1968. The Probable Function of Alligator Holes as Phosphate Traps Due to Periodic Drying of the Everglades. unpublished report, 5 pp.

131a. Lodge, Thomas E. (personal observations of the author).

131b. Lodge, Thomas E., Richard B. Darling, David J. Fall, and Hilburn O. Hillestad. 1994. A wetland evaluation method for the Everglades: impact to mitigation. Presented at the Florida Water Policy and Management Seminar, Telluride Colorado, January 20, 1994.

132. Loftin, Jan P. 1992. In the wake of Hurricane Andrew. *Florida Water* 1(2):2–8.

133. Loftus, William F. (personal communications). Fisheries Biologist, U.S. National Park Service, South Florida Research Center, Everglades National Park, Homestead, Florida.

134. Loftus, William F. and Oron Bass, Jr. 1992. Mercury threatens wildlife resources. *Park Science* 24(4):18–20.

134a. Loftus, William F. and Anne-Marie Eklund. 1994. Long-term dynamics of an Everglades small-fish assemblage. *In* Steven M. Davis and John C. Ogden (editors). *Everglades: The Ecosystem and Its Restoration*. St. Lucie Press, Delray Beach, Florida.

135. Loftus, William F., James D. Chapman, and Roxanne Conrow. 1986. Hydroperiod effects on Everglades marsh food webs, with relation to marsh restoration efforts. *In* Gary Larson and Michael Soukup (editors). Proceedings of the Conference on Science in the National Parks, Volume 6: Fisheries and Coastal Wetlands Research, pp. 1–22.

136. Loftus, William F., Robert A. Johnson, and Gordon H. Anderson. 1992. Ecological impacts of the reduction of groundwater levels in short-hydroperiod marshes of the Everglades. First International Conference on Ground Water Ecology, U.S. Environmental Protection Agency/American Water Resources Association, April 1992.

137. Loftus, William F. and James A. Kushlan. 1987. Freshwater fishes of southern Florida. *Bulletin Florida State Museum, Biological Sciences* 31(4):147–344.

137a. Logan, Todd, Andrew C. Eller, Jr., Ross Morrell, Donna Ruffner, and Jim Sewell. 1993. Florida Panther Habitat Preservation Plan, South Florida Population. Prepared by the U.S. Fish and Wildlife Service, the Florida Game and Fresh Water Fish Commission, the Florida Department of Environmental Protection, and the National Park Service for the Florida Panther Interagency Committee (approved January 3, 1994).

138. Long, Robert W. 1984. Origin of the vascular flora of southern Florida. *In* P.J. Gleason

(editor). *Environments of South Florida: Present and Past II*. Memoir II, Miami Geological Society, Miami, Florida.

139. Long, Robert W. and Olga Lakela. 1971. *A Flora of Tropical Florida*. University of Miami Press, Coral Gables, Florida.

140. Loveless, Charles M. 1959. A study of the vegetation in the Florida Everglades. *Ecology* 40(1):1–9.

141. Luer, Carlyle A. 1972. *The Native Orchids of Florida*. The New York Botanical Garden. W.S. Cowell Ltd., Ipswich, England, U.K.

142. MacArthur, R.H. and E.O. Wilson. 1967. *The Theory of Island Biogeography*. Princeton University Press, Princeton, New Jersey.

143. MacVicar, Thomas K. 1987. Rescuing the Everglades. *Civil Engineering* 57(8):40–42.

144. MacVicar, Thomas K. and Steve S.T. Lin. 1984. Historical rainfall activity in central and southern Florida: average, return period estimates and selected extremes. *In* P.J. Gleason (editor). *Environments of South Florida: Present and Past II*. Memoir II, Miami Geological Society, Miami, Florida.

145. Maehr, David S. (personal communications, 1989). Florida Game and Fresh Water Fish Commission, Naples, Florida.

146. Maffei, Mark D. 1991. Melaleuca control on Arthur R. Marshall Loxahatchee National Wildlife Refuge. *In* Ted D. Center, Robert F. Doren, Ronald H. Hofstetter, Ronald L. Myers, and Louis D. Whiteaker (editors). Proceedings of the Symposium on Exotic Pest Plants. U.S. Department of the Interior, National Park Service, Washington, D.C.

146a. Mairson, Alan. 1993. The Everglades: dying for help. *National Geographic* 185(4):2–35.

147. Marine, Gene. 1971. Algae and Aerojet. *In* William Ross McCluney (editor). *The Environmental Destruction of South Florida*. University of Miami Press, Coral Gables, Florida.

148. Mayr, Ernst. 1966. *Animal Species and Evolution*. The Belknap Press of Harvard University Press, Cambridge, Massachusetts.

149. Mazzotti, Frank J. 1989. Factors affecting the nesting success of the American crocodile, *Crocodylus acutus*, in Florida Bay. *Bulletin of Marine Science* 44(1):220–228.

149a. Mazzotti, Frank J. and Laura A Brandt. 1994. Ecology of the American alligator in a seasonally fluctuating environment. *In* Steven M. Davis and John C. Ogden (editors). *Everglades: The Ecosystem and Its Restoration*. St. Lucie Press, Delray Beach, Florida.

150. McClain, Michael and Enid Y. Karr. 1989. The Florida basement: diary of a turbulent past. *The Compass* 66(2):52–58.

151. McClintock, Jack. 1993. Crocodiles come back. *Sea Frontiers* 39(2):42–49.

152. McDiarmid, Roy W. (editor). 1978. *Volume Three: Amphibians and Reptiles*. *In* Peter C.H. Pritchard (series editor). *Rare and Endangered Biota of Florida*. University Presses of Florida, Gainesville, Florida.

152a. McIvor, Carole C., Janet A. Ley, and Robin D. Bjork. 1994. Changes in freshwater inflow from the Everglades to Florida Bay. *In* Steven M. Davis and John C. Ogden (editors). *Everglades: The Ecosystem and Its Restoration*. St. Lucie Press, Delray Beach, Florida.

153. McPherson, Benjamin F. 1984. The Big Cypress Swamp. *In* P.J. Gleason (editor). *Environments of South Florida: Present and Past II*. Memoir II, Miami Geological Society, Miami, Florida.

154. McPherson, Benjamin F., G.Y. Hendrix, H. Klein, and H.M. Tyus. 1976. The Environment of South Florida, A Summary Report. U.S. Geological Survey Professional Paper 1011.

155. Meeder, J.F. and L.B. Meeder. 1989. Hurricanes in Florida Bay: a dominant physical process (abstract). *Bulletin of Marine Science* 44(1):518.

156. Milanich, Jerald T. and Charles H. Fairbanks. 1980. *Florida Archaeology*. Academic Press, New York, New York.

157. Missimer, Thomas M. 1984. The geology of South Florida: a summary. *In* P.J. Gleason (editor). *Environments of South Florida: Present and Past II*. Memoir II, Miami Geological Society, Miami, Florida.

158. Mitsch, William J. and James G. Gosselink. 1993. *Wetlands* (second edition). Van Nostrand Reinhold, New York, New York.

159. Moler, Paul E. (editor). 1992. *Volume III: Amphibians and Reptiles. In* Ray E. Ashton, Jr. (series editor). *Rare and Endangered Biota of Florida*. University Presses of Florida, Gainesville, Florida.

160. Molnar, George, Ronald H. Hofstetter, Robert F. Doren, Louis D. Whiteaker, and Michael T. Brennan. 1991. Management of *Melaleuca quinquenervia* within East Everglades wetlands. *In* Ted D. Center, Robert F. Doren, Ronald H. Hofstetter, Ronald L. Myers, and Louis D. Whiteaker (editors). Proceedings of the Symposium on Exotic Pest Plants. U.S. Department of the Interior, National Park Service, Washington, D.C.

161. Monastersky, Richard. 1993. The long view of weather. *Science News* 144(21):328–330.

162. Murphy, J. Brendan and R. Damiam Nance. 1992. Mountain belts and the supercontinent cycle. *Scientific American* 266:84–91.

162a. National Audubon Society. 1994. Report on Water Supply Preserves. National Audubon Society Everglades System Campaign, Miami, Florida.

162b. National Geographic Society. 1987. *Field Guide to the Birds of North America* (second edition). National Geographic Society, Washington, D.C.

163. Neill, Wilfred T. 1969. *The Geography of Life*. Columbia University Press, New York, New York.

164. Newcott, William R. 1993. Lightning, nature's high-voltage spectacle. *National Geographic* 184(1):83–103.

165. Niering, William A. 1990. Vegetation dynamics in relation to wetland creation. *In* Jon A. Kusler and Mary E. Kentula (editors). *Wetland Creation and Restoration: The Status of the Science*. Island Press, Washington, D.C.

166. NOAA. 1980. Eastern United States Coastal and Ocean Zones Data Atlas. Prepared by G. Carleton Ray, M. Geraldine McCormick-Ray, James A. Dobbin. Charles N. Ehler, and Daniel J. Basta for the Council on Environmental Quality, Executive Office of the President, and the Office of Coastal Zone Management, National Oceanic and Atmospheric Administration, U.S. Department of Commerce, Washington, D.C.

167. Odum, Eugene P. 1971. *Fundamentals of Ecology* (third edition). W.B. Saunders, Philadelphia, Pennsylvania.

168. Odum, William E. and Carole C. McIvor. 1990. Mangroves. *In* Ronald L. Myers and John J. Ewel (editors). *Ecosystems of Florida*. University of Central Florida Press, Orlando, Florida.

169. Office of the Governor. 1992. Save Our Everglades. A status report by the Office of the Governor Lawton Chiles (Florida), January 24, 1992.

170. Ogden, John C. (personal communications). Biologist, U.S. National Park Service, South Florida Research Center, Everglades National Park, Homestead, Florida.

171. Ogden, John C. 1985. The wood stork. *Audubon Wildlife Report 1985*. The National Audubon Society, New York, New York.

171a. Ogden, John C. 1994. A comparison of wading bird nesting colony dynamics (1931–1946 and 1974–1989) as an indication of ecosystem conditions in the southern Everglades. *In* Steven M. Davis and John C. Ogden (editors). *Everglades: The Ecosystem and Its Restoration*. St. Lucie Press, Delray Beach, Florida.

172. Ogden, John C. and Robert A. Johnson. 1990. Ecosystem restoration in Everglades National Park: a prerequisite for wildlife recovery (draft). Presentation at the 56th North American Wildlife and Natural Resources Conference, Wildlife Management Institute, Washington, D.C., March 1991.

173. Ogden, John C., J.A. Kushlan, and J.T. Tilmant. 1978. The Food Habits and Nesting Success of Wood Storks in Everglades National Park, 1974. Natural Resources Report Number 16, National Park Service, U.S. Department of the Interior.

174. Ogden, John C. and Betsy Trent Thomas. 1985. A colonial wading bird survey in the central Llanos of Venezuela. *Colonial Waterbirds* 8(1):23–31.

175. Olmsted, Ingrid and Lloyd L. Loope. 1984. Plant communities of Everglades National Park. *In* P.J. Gleason (editor). *Environments of South Florida: Present and Past II.* Memoir II, Miami Geological Society, Miami, Florida.

176. Orians, Gordon H., Michael Bean, Russell Lande, Kent Loftin, Stuart Pimm, R. Eugene Turner, and Milton Weller. 1992. *Report of the Advisory Panel on the Everglades and Endangered Species.* Audubon Conservation Report No. 8, National Audubon Society, New York, New York.

177. Parker, Garald. 1984. Hydrology of the pre-drainage system of the Everglades in Southern Florida. *In* P.J. Gleason (editor). *Environments of South Florida: Present and Past II.* Memoir II, Miami Geological Society, Miami, Florida.

178. Paulson, D.R. 1966. The Dragonflies (Odonata: Anisoptera) of Southern Florida. Ph.D. dissertation, University of Miami, Coral Gables, Florida.

179. Payne, James F. 1978. Aspects of the life histories of selected species of North American crayfishes. *Fisheries* 3(6):5–8.

180. Pennak, Robert W. 1978. *Fresh-water Invertebrates of the United States* (second edition). John Wiley & Sons, New York, New York.

181. Penny, Malcolm. 1991. *Alligators and Crocodiles.* Crescent Books, New York, New York.

182. Perfit, Michael R. and Ernest E. Williams. 1989. Geological contraints [sic] and biological retrodictions in the evolution of the Caribbean Sea and its islands. *In* C.A. Woods (editor). *Biogeography of the West Indies, Past, Present, and Future.* Sandhill Crane Press, Gainesville, Florida.

183. Phillips, Walter S. 1940. A tropical hammock on the Miami (Florida) limestone. *Ecology* 21(2):166–175.

184. Pielou, E.C. 1991. *After the Ice Age.* University of Chicago Press, Chicago, Illinois.

185. Pirkle, E.C., W.H. Yoho, and C.W. Hendry, Jr. 1970. Ancient Sea Level Stands in Florida. Geological Bulletin No. 52, Florida Department of Natural Resources, Bureau of Geology, Tallahassee, Florida.

185a. Pollman, Curtis D. (personal communication). Vice President for Research and Development, KBN Engineering and Applied Sciences, Inc., Gainesville, Florida.

186. Pope, R. L., J.A. Browder, and P.B. Schroeder. 1980. A Study of the Feasibility of Using Stomach Contents Analyses to Determine Food Webs in the Everglades Periphyton Community. Contract PX52809, A report to Everglades National Park from the South Florida Environmental Research Foundation, Miami, Florida.

187. Powell, Allyn B., Donald E. Hoss, William F. Hettler, David S. Peters, and Stephanie Wagner. 1989. Abundance and distribution of ichthyoplankton in Florida Bay and adjacent waters. *Bulletin of Marine Science* 44(1):35–48.

188. Powell, George V.N., Robin D. Bjork, John C. Ogden, Richard T. Paul, A. Harriett Powell, and William B. Robertson, Jr. 1989. Population trends in some Florida Bay wading birds. *Wilson Bulletin* 101(3):436–457.

189. Pratt, Harry D. and Kent S. Littig. 1971. Mosquitos of Public Health Importance and Their Control. U.S. Department of Health, Education, and Welfare, Public Health Service, Health Services and Mental Health Administration, Bureau of Community Environmental Management, Atlanta, Georgia.

190. Rankin, Robert A., Steven Thomma, and Stephen K. Doig. 1988. Ominous warming: the greenhouse effect. *The Miami Herald*, October 16–18, 1988 (three-part series).

191. Reeder, Pamela B. and Steven M. Davis. 1983. Decomposition, Nutrient Uptake and Microbial Colonization of Sawgrass and Cattail Leaves in Water Conservation Area 2A. Technical Publication #83-4, South Florida Water Management District, West Palm Beach, Florida.

192. Rhoads, Peter B. (unpublished). Draft manuscript (July 1980) concerning crayfish (*Procambarus alleni*) reproduction and early growth stages in the southern Everglades and Big Cypress Swamp. Current address: South Florida Water Management District, West Palm Beach, Florida.

193. Rich, Earl R. (personal communications). Unpublished data concerning drought tolerance of the apple snail (*Pomacea paludosa*), and on snail dispersal in the Everglades. Professor of Biology Emeritus, Department of Biology, University of Miami, Coral Gables, Florida.

193a. Richardson, John R., Wade L. Bryant, Wiley M. Kitchens, Jennifer E. Mattson, and Kevin R. Pope. 1990. An Evaluation of Refuge Habitats and Relationship to Water Quality, Quantity and Hydroperiod, A Synthesis Report. Prepared for the Arthur R. Marshall Loxahatchee National Wildlife Refuge, Boynton Beach, Florida by the Florida Cooperative Fish and Wildlife Research Unit, University of Florida, Gainesville, Florida.

194. Robbin, Daniel M. 1984. A new Holocene sea level curve for the upper Florida Keys and Florida reef tract. *In* P.J. Gleason (editor). *Environments of South Florida: Present and Past II*. Memoir II, Miami Geological Society, Miami, Florida.

195. Robblee, M.B., T.R. Barber, P.R. Carlson, Jr., M.J. Durako, J.W. Fourqurean, L.K. Muehlstein, D. Porter, L.A. Yarbro, R.T. Zieman, and J.C. Zieman. 1991. Mass mortality of the tropical seagrass *Thalassia testudinum* in Florida Bay (USA). *Marine Ecology Progress Series* 71: 297–299.

196. Robertson, William B., Jr. (personal communication). U.S. National Park Service, South Florida Research Center, Everglades National Park, Homestead, Florida.

197. Robertson, William B., Jr. 1954. Everglades fires—past, present and future. *Everglades Natural History* 2(1):10–16.

198. Robertson, William B., Jr. 1989. *Everglades: The Park Story* (revised edition). Florida National Parks and Monuments Association, Homestead, Florida.

199. Robertson, William B., Jr. 1993. Oral address at the Everglades Coalition 8th Annual Conference, opening session, Tallahassee, Florida, February 20, 1993.

199a. Robertson, William B., Jr. and Peter C. Frederick. 1994. The faunal chapters: contexts, synthesis, and departures. *In* Steven M. Davis and John C. Ogden (editors). *Everglades: The Ecosystem and Its Restoration*. St. Lucie Press, Delray Beach, Florida.

200. Robertson, William B., Jr. and James A Kushlan. 1984. The southern Florida avifauna. *In* P.J. Gleason (editor). *Environments of South Florida: Present and Past II*. Memoir II, Miami Geological Society, Miami, Florida.

201. Robins, C. Richard (personal communication). Maytag Professor of Ichthyology, Rosenstiel School of Marine and Atmospheric Science, University of Miami, Miami, Florida.

202. Robins, C. Richard and G. Carleton Ray. 1986. *A Field Guide to Atlantic Coast Fishes of North America*. Houghton Mifflin, Boston, Massachusetts.

203. Robins, C. Richard, Reeve M. Bailey, Carl E. Bond, James R. Brooker, Ernest A. Lachner, Robert N. Lea, and W.B. Scott. 1991. *Common and Scientific Names of Fishes from the United States and Canada* (revised edition). Special Publication 20, American Fisheries Society, Bethesda, Maryland.

204. Rosendahl, P.C. and P.W. Rose. 1982. Freshwater flow rates and distribution within the Everglades marsh. *In* R.D. Cross and D.L. Williams (editors). Proceedings of the National Symposium on Freshwater Inflow to Estuaries, Coastal Ecosystems Project. U.S. Fish and Wildlife Service, Washington, D.C.

205. Ross, Charles A. (editor). 1989. *Crocodiles and Alligators*. Facts on File, New York, New York.

206. Runde, D.E. 1991. Trends in Wading Bird Nesting Populations in Florida: 1976–1978 BS 1986–1989 (Final Performance Report). Florida Game and Fresh Water Fish Commission Nongame Wildlife Program, 90 pp.

207. Rutherford, Edward S., James T. Tilmant, Edith B. Thue, and Thomas W. Schmidt. 1989. Fishery harvest and population dynamics of spotted seatrout, *Cynoscion nebulosus*, in Florida bay and adjacent waters. *Bulletin of Marine Science* 44(1):108–125.

207a. Schemnitz, S.D. and J.L. Schortemeyer. 1973. The impact of half tracks and airboats on the Florida Everglades environment. *In* Proceedings of the 1973 Snowmobile and

Off-the-Road-Vehicle Research Symposium. Technical Report No. 9, Michigan State University, East Lansing, Michigan.

208. Schneider, William J. and James H. Hartwell. 1984. Troubled waters of the Everglades. *Natural History* 93(11):46–57.

209. Schokman, Larry (supervisor). 1985. *Plants of the Kampong*. Harvard University Printing Office, Allston, Massachusetts.

210. Scott, Gerald P., Michael R. Dewey, Larry J. Hansen, Ralph E. Owen, and Edward S. Rutherford. 1989. How many mullet are there in Florida Bay? *Bulletin of Marine Science* 44(1):89–107.

211. SFWMD. 1992. Surface Water Improvement and Management Plan for the Everglades—Supporting Information Document (March 13, 1992). South Florida Water Management District, West Palm Beach, Florida.

212. SFWMD. 1993. Everglades Restoration Annual Progress Report January 1993. South Florida Water Management District, West Palm Beach, Florida.

213. SFWMD. 1993. Focus on Florida Bay. *Everglades Connection* 2(1):1 and 3–5.

213a. SFWMD. 1993. Lower East Coast Regional Water Supply Plan, Appendices and Technical Information, Draft Working Document. South Florida Water Management District Planning Department, West Palm Beach, Florida.

213b. SFWMD. 1994. Everglades Restoration Annual Progress Report January 1994. South Florida Water Management District, West Palm Beach, Florida.

214. Smith, Douglas L. 1982. Review of the tectonic history of the Florida basement. *Tectonophysics* 88:1–22.

215. Smith, Joel B. 1990. From global to regional climate change: relative knowns and unknowns about global warming. *Fisheries* 15(6):2–6.

216. Smith, Thomas J., III, J. Harold Hudson, Michael B. Robblee, George V. N. Powell, and Peter J. Isdale. 1989. Freshwater flow from the Everglades to Florida Bay: a historical reconstruction based on fluorescent banding in the coral *Solenastrea bournoni*. *Bulletin of Marine Science* 44(1):274–282.

216a. Snyder, G.H. and J.M. Davidson. 1994. Everglades agriculture: past, present, and future. *In* Steven M. Davis and John C. Ogden (editors). *Everglades: The Ecosystem and Its Restoration*. St. Lucie Press, Delray Beach, Florida.

217. Snyder, James R., Alan Herndon, and William B. Robertson, Jr. 1990. South Florida rockland. *In* Ronald L. Myers and John J. Ewel (editors). *Ecosystems of Florida*. University of Central Florida Press, Orlando.

218. Sogard, Susan M., George V.N. Powell, and Jeff G. Holmquist. 1989. Spatial distribution and trends in abundance of fishes residing in seagrass meadows on Florida Bay mudbanks. *Bulletin of Marine Science* 44(1):179–199.

218a. Spalding, Marilyn G. (personal communication). Department of Infectious Diseases, College of Veterinary Medicine, University of Florida, Gainesville, Florida.

219. Stafford, W.E. 1919. Natural history of Paradise Key and the near-by Everglades of Florida. *In* The Annual Report of the Board of Regents of the Institution for the Year Ending June 30, 1917. Smithsonian Institution, Washington, D.C.

220. Stearn, Colin W., Robert L. Carroll, and Thomas H. Clark. 1979. *Geological Evolution of North America*. John Wiley & Sons, New York, New York.

221. Stephens, John C. 1984. Subsidence of organic soils in the Florida Everglades—a review and update. *In* P.J. Gleason (editor). *Environments of South Florida: Present and Past II*. Memoir II, Miami Geological Society, Miami, Florida.

222. Sun-Sentinel staff. 1992. *Andrew! Savagery from the Sea*. Tribune Publishing, Orlando, Florida.

223. Sweeney, Catherine H. and Harriet S. Fraunfelter (personal communications). The Kampong, Coconut Grove, Florida.

224. Swift, David R. 1984. Periphyton and water quality relationships in the Everglades water

conservation areas. *In* P.J. Gleason (editor). *Environments of South Florida: Present and Past II*. Memoir II, Miami Geological Society, Miami, Florida.

225. Tabb, Durbin C. and Martin A. Roessler. 1989. History of studies on juvenile fishes of coastal waters of Everglades National Park. *Bulletin of Marine Science* 44(1):23–34.

226. Tauvers, Peter R. and William R. Muehlberger. 1987. Is the Brunswick magnetic anomaly really the alleghanian suture? *Tectonics* 6(3):331–342.

227. Taylor, William Randolph. 1960. *Marine Algae of the Eastern Tropical and Subtropical Coasts of the Americas*. The University of Michigan Press, Ann Arbor, Michigan.

228. Tebeau, Charlton W. 1968. *Man in the Everglades*. University of Miami Press, Coral Gables, Florida.

229. Tebeau, Charlton W. 1984. Exploration and early descriptions of the Everglades, Lake Okeechobee, and the Kissimmee River. *In* P.J. Gleason (editor). *Environments of South Florida: Present and Past II*. Memoir II, Miami Geological Society, Miami, Florida.

230. Thomas, Terence M. 1974. A detailed analysis of climatological and hydrological records of South Florida with reference to man's influence upon the ecosystem evolution. *In* P.J. Gleason (editor). *Environments of South Florida: Present and Past*. Memoir 2, Miami Geological Society, Miami, Florida.

231. Thompson, Fred G. 1984. *Freshwater Snails of Florida*. University of Florida Press, Gainesville, Florida.

232. Thompson, Fred G. (personal communication). Department of Natural Sciences, Florida Museum of Natural History, Gainesville, Florida.

233. Tilmant, James T. 1989. A history and an overview of recent trends in fisheries of Florida Bay. *Bulletin of Marine Science* 44(1):3–22.

234. Tilmant, James T., Edward S. Rutherford, and Edith B. Thue. 1989. Fishery harvest and population dynamics of red drum (*Sciaenops ocellatus*) from Florida Bay and adjacent waters. *Bulletin of Marine Science* 44(1):126–138.

235. Timmer, C. Elroy and Stanley S. Teague. 1991. Melaleuca eradication program: assessment of methodology and efficacy. *In* Ted D. Center, Robert F. Doren, Ronald H. Hofstetter, Ronald L. Myers, and Louis D. Whiteaker (editors). Proceedings of the Symposium on Exotic Pest Plants. U.S. Department of the Interior, National Park Service, Washington, D.C.

236. Tomlinson, P.B. 1980. *The Biology of Trees Native to Tropical Florida*. Harvard University Printing Office, Allston, Massachusetts.

237. Tomlinson, P.B. 1986. *The Botany of Mangroves*. Cambridge University Press, London.

238. Toops, Connie M. 1979. *The Alligator: Monarch of the Everglades*. The Everglades Natural History Association, Homestead, Florida.

239. Toth, Louis A. 1987. Effects of Hydrologic Regimes on Lifetime Production and Nutrient Dynamics of Sawgrass. Technical Publication #87-6, South Florida Water Management District, West Palm Beach, Florida.

240. Turgeon, Donna D., Arthur E. Bogan, Eugene V. Coan, William K. Emerson, William G. Lyon, William L. Pratt, Clyde F.E. Roper, Amelie Scheltema, Fred G. Thompson, and James D. Williams. 1988. *Common and Scientific Names of Aquatic Invertebrates from the United States and Canada: Mollusks*. Special Publication 16, American Fisheries Society, Bethesda, Maryland.

241. U.S. Army Corps of Engineers. (undated [1981?]). An Environmental Assessment Related to the Construction, Use and Maintenance of the Okeechobee Waterway, Florida. Technical Report, Jacksonville District, Contract No. DACW17-80-C-0048.

242. U.S. Fish and Wildlife Service. 1987. Florida Panther (*Felis concolor coryi*) Recovery Plan. Prepared by the Florida Panther Interagency Committee for the U.S. Fish and Wildlife Service, Atlanta, Georgia.

243. VanArman, Joel. 1984. South Florida's estuaries. *In* P.J. Gleason (editor). *Environments of South Florida: Present and Past II*. Memoir II, Miami Geological Society, Miami, Florida.

244. Van Meter, Victoria Brook. 1985. *Florida's Wood Storks*. Florida Power and Light Company, Miami, Florida.

245. Wade, Dale, John Ewel, and Ronald Hofstetter. 1980. Fire in South Florida Ecosystems. Forest Service Technical Report SE-17, U.S. Dept of Agriculture, Southeastern Forest Experiment Station, Asheville, North Carolina.

246. Waller, Bradley G. 1988. Hydrologic effects of drainage and water management on the wetland ecosystem of South Florida (U.S.A.). *In* Miguel Chávez Miguel (editor). *Ecología y Conservación del Delta de los ríos Usumacinta y Grijalva* (Memorias). Instituto Nacional de Investigacíon sobre Recursos Bióticos (INIREB)—División Regional Tabasco, pp. 201–215 (see Reference 30).

247. Walsh, Gerald E. 1974. Mangroves: a review. *In* Robert J. Reimold and William H Queen (editors). *Ecology of Holophytes*. Academic Press, New York, New York.

247a. Walters, Carl J. and Lance H. Gunderson. 1994. A screening of water policy alternatives for ecological restoration in the Everglades. *In* Steven M. Davis and John C. Ogden (editors). *Everglades: The Ecosystem and Its Restoration*. St. Lucie Press, Delray Beach, Florida.

248. Walters, Carl, Lance Gunderson, and C.S. Holling. 1992. Experimental polices for water management in the Everglades. *Ecological Applications* 2(2):189–202.

249. Wanless, Harold R. (personal communication). Professor of Marine Geology, Rosenstiel School of Marine and Atmospheric Science, University of Miami, Miami, Florida.

250. Wanless, Harold R. 1984. Mangrove sedimentation in geologic perspective. *In* P.J. Gleason (editor). *Environments of South Florida: Present and Past II*. Memoir II, Miami Geological Society, Miami, Florida.

251. Wanless, Harold R. 1989. The inundation of our coastlines (past, present and future with a focus on South Florida). *Sea Frontiers* 35(5):264–271.

252. Wanless, Harold R. and Matthew G. Tagett. 1989. Origin, growth and evolution of carbonate mudbanks in Florida Bay. *Bulletin of Marine Science* 44(1):490–514.

252a. Wanless, Harold R., Randall W. Parkinson, and Lenore P. Tedesco. 1994. Sea level control on stability of Everglades wetlands. *In* Steven M. Davis and John C. Ogden (editors). *Everglades: The Ecosystem and Its Restoration*. St. Lucie Press, Delray Beach, Florida.

253. Ward, Daniel B. (editor). 1979. *Volume Five: Plants*. *In* Peter C.H. Pritchard (series editor). *Rare and Endangered Biota of Florida*. University Presses of Florida, Gainesville, Florida.

254. Ware, Forrest J., Homer Royals, and Ted Lange. 1990. Mercury contamination in Florida largemouth bass. *Proceedings of the Annual Conference of the Southeastern Association of Fish and Wildlife Agencies* 44:5–12.

255. Weaver, James, Bradford Brown, Joan Browder, Wiley Kitchens, David Wesley, David Unsell, Lawrence Burns, Dan Scheidt, David Ferrell, Nancy Thompson, John Ogden, Tom Armentano, Michael Robblee, William Loftus, Barry Glaz, and Peter Ortner. 1993. "Federal Objectives for the South Florida Restoration" by the Science Sub-Group of the South Florida Management and Coordination Working Group established by the Interagency Agreement on South Florida Ecosystem Restoration, September 23, 1993, between the United States Departments of Interior, Commerce, Army (Civil Works), Justice, and Agriculture, and the Environmental Protection Agency.

256. Webb, S. David. 1990. Historical biogeography. *In* Ronald L. Myers and John J. Ewel (editors). *Ecosystems of Florida*. University of Central Florida Press, Orlando.

256a. White, Peter S. 1994. Synthesis: vegetation pattern and process in the Everglades ecosystem. *In* Steven M. Davis and John C. Ogden (editors). *Everglades: The Ecosystem and Its Restoration*. St. Lucie Press, Delray Beach, Florida.

257. Williams, Austin B., Lawrence G. Abele, Darryl L. Felder, Horton H. Hobbs, Jr., Raymond B. Manning, Patsy A. McLaughlin, and Isabel Pérez Farfante. 1989. *Common and Scientific Names of Aquatic Invertebrates from the United States and Canada: Decapod Crustaceans*. Special Publication 16, American Fisheries Society, Bethesda, Maryland.

258. Willoughby, Hugh L. 1898. *Across the Everglades, A Canoe Journey of Exploration* (fifth edition). Florida Classics Library Edition (1992), Port Salerno, Florida.

259. Windley, Brian F. 1984. *The Evolving Continents* (second edition). John Wiley & Sons, New York, New York.

259a. Wood, Don A. (compiler). 1993. Official Lists of Endangered & Potentially Endangered Fauna and Flora in Florida (October 1, 1993). Florida Game and Fresh Water Fish Commission, Tallahassee, Florida.

260. Wright, Myron H., Jr., Ray O. Green, Jr., and Norman D. Reed. 1972. The swallow-tailed kite: graceful aerialist of the Everglades. *National Geographic* 142(4):496–505.

261. Yates, S.A. 1974. An Autecological Study of Sawgrass, *Cladium jamaicense*, in Southern Florida. Masters thesis, Department of Biology, University of Miami, Coral Gables, Florida.

262. Zolczynski, Stephen J., Jr. and William D. Davis. 1976. Growth characteristics of the northern and Florida subspecies of largemouth bass and their hybrid, and a comparison of catchability between the subspecies. *Transactions of the American Fisheries Society* 105(2):240–243.

263. Zuckerman, Bertram. 1993. *The Kampong, The Fairchilds' Tropical Paradise*. National Tropical Botanical Garden and Fairchild Tropical Garden.

INDEX

Acer rubrum, 40, see also Red maple
Acris gryllus dorsalis, 124
Acrostichum danaeifolium, 40, see also Leather fern
Adventitious roots, 65, see also Prop roots
Aedes taeniorhynchus, 71, see also Salt-marsh mosquito
Aerojet Canal, 183
Aestivation
  alligator, 136–137, 180
  bowfin, 113
  Florida applesnail, 100–101
  Florida tree snail, 106–107
Agkistrodon piscivorus conanti, 128, see also Florida cottonmouth
Agriculture, see also Sugarcane
  marsh soils and, 34
  pond apple swamp and, 45
  restoration and, 172–173, 185, 187
  sawgrass marsh and, 23
  water quality and, 180
Airboat, 168, 169, 170
Air plants, 94
  hammocks and, 51
  hurricanes and, 77, 79
  pond apple and, 46
Ajaia ajaja, 149, see also Roseate spoonbill
Albula vulpes, 119, see also Bonefish
Algae, 23, 30–33, 103, see also Periphyton
  amphibians and, 126
  blue-green, see Blue-green algae
  Florida Bay and, 83, 84, 85, 183
  green, see Green algae
  marl and, 33
  riverine grass shrimp and, 104
  Seminole rams-horn and, 102
Algal mat, 31, 101, see also Periphyton
Alleghanian suture, 5

Alligator, American, 127, 128, 132–137, see also Alligator hole
  bay heads and, 42
  birds and, 154
  compared to other crocodilians, 132–133
  danger to man, 132, 135
  distribution, 134
  Florida panther and, 143
  freshwater fishes and, 109, 111, 112
  habitats, 186
  importance, 132, 135–137
  invertebrates and, 101, 103–104, 107
  life cycle, 134–135
  mangrove swamp and, 66
  mercury and, 181
  in ponds and creeks, 30
  predation, 129, 133
  predators of, 133
  protection of, 134, 176
  reproduction, 135
  restoration and, 179–180, 186, 189
  sawgrass marsh and, 24, 25
  size, 132
  territoriality, 135
  trails, 135, 136
  wading birds and, 155
Alligator, Chinese, 133
Alligator flag, see Flag
Alligator gar, 112
Alligator hole, 22, 30, 31, 135–137
  bay heads and, 42, 43
  drought and, 30
  pond apple and, 45
  restoration and, 179, 180
  riverine grass shrimp and, 104
  soils in, 35
  tree islands and, 47
  willow and, 42
Alligator mississippiensis, 128, see also Alligator, American

211

Y

Z